# From Materials to Applications: High-Performance Steel Structures

# From Materials to Applications: High-Performance Steel Structures

Guest Editor

**Alireza Farzampour**

MDPI

Basel • Beijing • Wuhan • Barcelona • Belgrade • Novi Sad • Cluj • Manchester

*Guest Editor*
Alireza Farzampour
Department of Permitting,
Inspections and Enforcement
Prince George County
Permitting & Licensing
Upper Marlboro
USA

*Editorial Office*
MDPI AG
Grosspeteranlage 5
4052 Basel, Switzerland

This is a reprint of the Special Issue, published open access by the journal *Materials* (ISSN 1996-1944), freely accessible at: https://www.mdpi.com/journal/materials/special_issues/4OA0A819PO.

For citation purposes, cite each article independently as indicated on the article page online and as indicated below:

| |
|---|
| Lastname, A.A.; Lastname, B.B. Article Title. *Journal Name* **Year**, *Volume Number*, Page Range. |

**ISBN 978-3-7258-8235-9 (Hbk)**
**ISBN 978-3-7258-8236-6 (PDF)**
**https://doi.org/10.3390/books978-3-7258-8236-6**

# Contents

**About the Editor** . . . . . . . . . . . . . . . . . . . . . . . . . . . . . . . . . . . . . . . . . . . . . . . . . **vii**

**Preface** . . . . . . . . . . . . . . . . . . . . . . . . . . . . . . . . . . . . . . . . . . . . . . . . . . . . . . . . . **ix**

**Yu Tao, Sumei Zhang, Gaopeng Xiong, Chao Gong, Zhaoxin Hou and Xiaozhong Li**
Eccentric Compression Behavior of Truss-Reinforced Cross-Shaped Concrete-Filled Steel Tubular Columns
Reprinted from: *Materials* **2024**, *17*, 3738, https://doi.org/10.3390/ma17153738 . . . . . . . . . . **1**

**Zhenyu Zhao, Jian Sheng, Dazhao Li, Shaobin Bai, Yongan Chen, Haitao Lu, et al.**
Achieving Strength–Ductility Balance in TWIP Steel by Tailoring Cementite
Reprinted from: *Materials* **2025**, *18*, 843, https://doi.org/10.3390/ma18040843 . . . . . . . . . . . **22**

**Mingjing Chang, Bo Chen, Xiang Xiao and Yanzhou Chen**
Nonstationary Stochastic Responses of Transmission Tower-Line System with Viscoelastic Material Dampers Under Seismic Excitations
Reprinted from: *Materials* **2025**, *18*, 1138, https://doi.org/10.3390/ma18051138 . . . . . . . . . . **38**

**Shun Bao, Han Feng, Zhigang Song, Jianguo He, Xiaohan Wu and Yang Gu**
Deformation Behavior of S32750 Duplex Stainless Steel Based on In Situ EBSD Technology
Reprinted from: *Materials* **2025**, *18*, 2030, https://doi.org/10.3390/ma18092030 . . . . . . . . . . **62**

**Qusai A. Alomari, Tewodros Y. Yosef, Robert W. Bielenberg, Ronald K. Faller, Mehrdad Negahban, Zesheng Zhang, et al.**
Material Characterization and Stress-State-Dependent Failure Criteria of AASHTO M180 Guardrail Steel: Experimental and Numerical Investigation
Reprinted from: *Materials* **2025**, *18*, 2523, https://doi.org/10.3390/ma18112523 . . . . . . . . . . **73**

**Feng Huang, Zhe Cheng, Defa Li, Wei Zhang and Zhili Hu**
Grain Refinement and Microstructural Evolution in Cobalt-Saving 18Ni (300) Maraging Steel via Cold Deformation-Cyclic Solution Treatment
Reprinted from: *Materials* **2025**, *18*, 2947, https://doi.org/10.3390/ma18132947 . . . . . . . . . . **105**

**Yinghu Wang, Long Chen, Limei Cheng, Enuo Wang, Zhendong Sheng and Ligang Zhang**
Machine Learning-Based Multi-Objective Composition Optimization of High Nitrogen Austenitic Stainless Steels
Reprinted from: *Materials* **2025**, *18*, 5460, https://doi.org/10.3390/ma18235460 . . . . . . . . . . **125**

**Yuhe Huang, Shuize Wang, Jiahao Qiang, Hui Wang and Jun Lu**
Effect of Si on Marine Corrosion Behavior of Austenite Low-Density Steel
Reprinted from: *Materials* **2026**, *19*, 1182, https://doi.org/10.3390/ma19061182 . . . . . . . . . . **146**

# About the Editor

**Alireza Farzampour**

Alireza Farzampour received his B.Sc. and M.Sc. degrees from Sharif University of Technology and worked as a research assistant in the field of materials. Following that, he received his PhD from Virginia Tech University. Dr. Farzampour has experience working at Virginia Tech University as a researcher and instructor of engineering university courses. Subsequently, he worked in the field of bridge engineering in the industry and for more than five years as a researcher and consultant infrastructure engineer in the United States. Having received his professional civil engineering license from the United States Board of Professional Engineers, he has been an active member of the American Board of Civil Engineers. His research interests include computational and experimental investigations of seismic resisting systems, innovative seismic steel protective structures, performance-based design and optimization, control and identification of structures, steel metallurgy, ductility, machine learning investigations, and brittle and ductile behavior.

# Preface

Advances in civil infrastructure systems that improve infrastructure resiliency against various environmental conditions are important for their ability to reduce damage, improve performance, and enhance the reliability of structures. Infrastructure can be designed, optimized, and constructed to withstand various lateral loads without experiencing a significant amount of concentrated damage. Having sufficient strength and stiffness with optimized materials to reduce the vulnerabilities to serious damage under various loading conditions requires structural elements to have adequate ductility, energy-dissipating capability, stiffness, and durability.

Several profound studies are collected in this Special Issue, aiming to investigate the advancements in high-performance steel structures: from materials to applications for use in various possible fields of study, which include the performance evaluations of steel structures, optimized steel systems, high-temperature reactions, innovative steel lateral resisting systems, high-strength steels, steel rebars, fatigue analysis methodologies, new generations of lateral resisting systems, natural hazard protective systems, additively manufactured steels, corrosion, application possibilities of new sustainable materials and methods, steel dampers, temporary structures, steel applications, steel–wood interactions, steel–concrete interactions, and advanced material behavior compared with steel. These subjects are generally studied from various points of view, including, but not limited to, mechanical, microstructural, material modeling, steel design, experimental and computational performance analyses and recent code compliance.

**Alireza Farzampour**
*Guest Editor*

*materials*

MDPI

*Article*

# Eccentric Compression Behavior of Truss-Reinforced Cross-Shaped Concrete-Filled Steel Tubular Columns

**Yu Tao [1,2], Sumei Zhang [1,2,*], Gaopeng Xiong [3], Chao Gong [3], Zhaoxin Hou [3] and Xiaozhong Li [1,2]**

[1] School of Civil and Environmental Engineering, Harbin Institute of Technology, University Town, Shenzhen 518055, China; 19b954014@stu.hit.edu.cn (Y.T.)
[2] Guangdong Provincial Key Laboratory of Intelligent and Resilient Structures for Civil Engineering, Harbin Institute of Technology, Shenzhen 518055, China
[3] Central Research Institute of Building and Construction Co., Ltd., MCC Group, Shenzhen 518055, China
* Correspondence: smzhang@hit.edu.cn

**Abstract:** In the paper, the eccentric compression behavior of the truss-reinforced cross-shaped concrete-filled steel tubular (CCFST) column is investigated. A total of eighteen CCFST columns were tested under eccentric compression, and the key test variables included the reinforced truss node spacing ($s$ = 140 mm and 200 mm), slenderness ratio ($\lambda$ = 9.2, 16.6, and 23.1), and eccentricity ratio ($\eta$ = 0, 0.08, and 0.15). The failure mode, deformation characteristic, stress distribution, strain distribution at the mid-span of the steel tube, and the eccentric compression bearing capacity were assessed. The results show that due to the addition of reinforced truss, the steel plates near the mid-span of eccentrically compressed CCFST columns experienced multi-wave buckling rather than single-wave buckling after the peak load was reduced to 85%, and the failure mode of concrete also changed from single-section to multi-section collapse failure. Comparisons were made with the unstiffened specimen. The ductility coefficient of the stiffened specimen with eccentricity ratios of 0.08–0.15 and node spacings of 140 mm~200 mm increased by 70~83%, approaching that of the multi-cell specimens with an increasing steel ratio of 1.8%. In addition, by comparing the test results with the calculation results of four domestic and international design codes, it was found that the Chinese codes CECS159-2018 and GB50936-2014, and the Eurocode 4 (2004) can be better employed to predict the compression bearing capacity of truss-reinforced CCFST columns.

**Keywords:** CCFST; reinforced truss; eccentric compression behavior; ductility; bearing capacity

## 1. Introduction

Special-shaped concrete-filled steel tubular (CFST) columns are usually designed in the form of T-shaped, cross-shaped, and L-shaped sections, whose column limbs are aligned with the walls, avoiding the exposure of indoor column corners, facilitating the placement of furniture, improving the efficiency of indoor space utilization, and meeting the functional requirements of residential buildings, which has attracted extensive attention in the engineering field and academia [1–7]. However, compared with circular or square or rectangular columns [8–11], traditional special-shaped CFST columns with concave corners tend to cause the separation of the materials, steel tube and concrete, under load. The separation prevents the steel tube from providing effective horizontal confinement to the concrete, thereby undermining the optimal performance of the steel tube and concrete.

To enhance the combined effect of steel tube and core concrete, scholars have proposed a series of methods to stiffen special-shaped CFST columns with the aim of increasing bearing capacity or ductility. One effective approach involves configuring multi-cell sections [2,3,7,12–16]. For instance, Du Guofeng et al. [17] conducted tests on the T-shaped multi-cell CFST members under eccentric compression. The results revealed that the multi-cell arrangement of the T-shaped CFST column could enhance synergistic working, leading to improved mechanical performance. Tu Yongqing et al. [14,15,18] conducted compression tests on T-shaped multi-cell CFST columns. Through the analysis of test data and finite element simulation results, the influence of various parameters on the column's bearing capacity was examined. Li Quan et al.'s [19] research on eccentric compression tests of T-shaped CFST columns with different slenderness ratios demonstrated that the impact of eccentricity directions (0° and 90°) was relatively minor. Some scholars have also improved the compression behavior of special-shaped CFST columns by adding longitudinal stiffening ribs [3,20–23], binding bars [24–30], or tensile bars [31]. For instance, scholars like Zuo Zhiliang, Cai Jian, and Long Yueling [24–28] conducted axial and eccentric compression tests on L-shaped and T-shaped CFST columns with binding bars. They analyzed the impact of horizontal spacings, eccentricity, and load angle on axial and eccentric compression behavior. Yang Yuanlong et al. [23,31] investigated T-shaped CFST columns with longitudinal stiffening ribs and tensile bars stiffener. They conducted axial and eccentric compression tests and conducted finite element analysis to study their bearing capacity and stability. The measures can enhance the synergistic working between the two materials, steel tube and concrete; however, several issues remain, such as more complex processing technology, steel plate sidewall unevenness, an insufficient improvement of component ductility, and a high steel ratio.

To address the above issues, a truss-reinforced CCFST column is proposed [32]. The design ingeniously welds bent bars to the edges of the grooved steel plate, forming reinforced truss within the specimen. The configuration better improves the restraining effect on the sidewalls of the steel plate. The reinforced truss form is characterized by high machining accuracy, high quality welding, and modular production after adopting automated equipment. Meanwhile, the tensile force on the diagonal steel bars is direct, and the effect on the steel plate at the concave corner is obvious. Compared with the multi-cell form, the steel consumption inside the member with reinforced truss is greatly reduced; compared with the binding bar form, the bent steel bars are only welded on the short steel plate extending inside the steel tube, which avoids the problem of opening holes on the plate and unevenness on the outer surface of the steel plate caused by bolts, and at the same time, it avoids the residual stresses on the steel tube generated by multiple welding. Another study [33] investigated the axial compressive mechanical properties of the truss-reinforced CCFST short column, and the results showed that the setting of reinforced trusses can effectively enhance the confinement effect of steel tube on concrete and the ductility of the member was significantly improved. However, the type of member has not been studied under eccentric compression. Therefore, this paper takes the truss-reinforced CCFST column as the research object, and a total of eighteen CCFST columns are tested under eccentric compression. The influence of parameters such as the reinforced truss node spacing, reinforced truss node spacing ($s$ = 140 mm and 200 mm), slenderness ratio ($\lambda$ = 9.2, 16.6, and 23.1), and eccentricity ratio ($\eta$ = 0, 0.08, and 0.15) on the member's eccentric compression behavior is investigated. The effect on the amount of eccentric compression behavior is also elaborated on the member's working mechanism under eccentric compression, and relevant suggestions on the calculation of the eccentric compression bearing capacity applicable to the truss-reinforced members are given, aiming at providing references for theoretical analyses and practical engineering applications.

## 2. Test Program

### 2.1. Details of Specimens

In order to study the eccentric compression behavior of truss-reinforced CCFST columns, 12 truss-reinforced CCFST specimens, 2 unstiffened comparative specimens, 2 spiral stirrup-stiffened comparative specimens, and 2 multi-cell comparative specimens were designed and fabricated, and the configuration forms are displayed in Figure 1. The limb height and limb thickness of all specimens were 120 mm. Concrete with a strength grade of C50 was used, the steel tube was made of steel plate with a thickness of 4 mm and a grade of Q235, and the truss was made of HPB300 steel bar with a diameter of 8 mm (Figure 2). Table 1 lists the parameters of specimens in detail; $\alpha$ is the steel ratio; CR and C in the specimen label represent the truss-reinforced specimen and the unstiffened specimen, respectively; and CL and CD represent the truss-reinforced cross-shaped specimen and the multi-cell specimen, respectively. The letters a, b, and c correspond to different slenderness ratios $\lambda$. The last group of numbers in all specimen labels represents the loading eccentricity $e$, and the reciprocal second group of numbers denotes the reinforced truss node spacing $s$ or the spiral stirrup pitch $d$. Taking "CRa-140-27" as an example, "CR" means that the specimen is a truss-reinforced specimen, "a" means that the slenderness ratio $\lambda$ is 9.2, "140" means that the reinforced truss node spacing $s$ is 140 mm, and "27" means that the load eccentricity $e$ is 27 mm, i.e., the eccentricity ratio is 0.08.

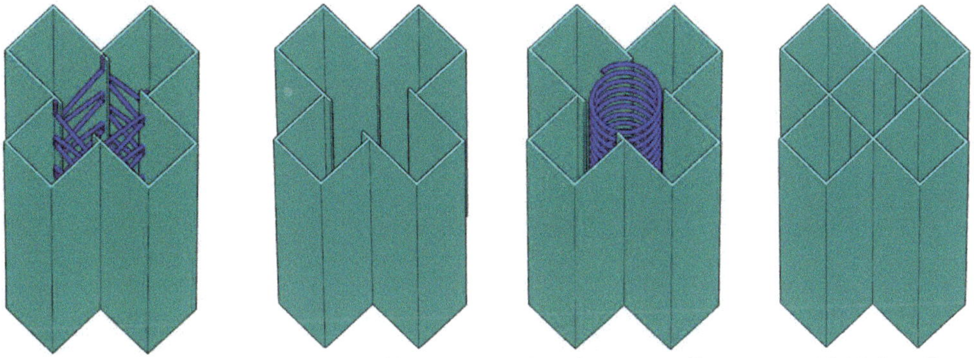

(**a**) Reinforced-truss stiffening (**b**) No stiffening (**c**) Spiral stirrup stiffening (**d**) Multi-cell

**Figure 1.** Schematic diagram of specimen structure.

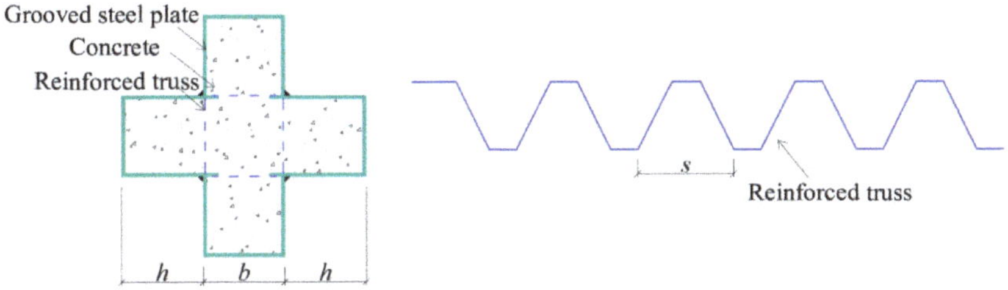


(**a**) Cross-section (**b**) Diagram of reinforced truss

**Figure 2.** The truss-reinforced CCFST column.

The fabrication of the specimen was carried out according to the configuration form of the designed truss-reinforced cross-shaped specimens. Firstly, the steel plates were cold-bent into groove shape, the steel bars were bent, and the bent bars were welded to the edges of the grooved steel plate (Figure 3a). The welding of steel plate and steel bar was carried out by $CO_2$ gas arc welding with a welding foot height of 5 mm, and continuous welding was carried out within the length (40 mm) of the short steel bar in contact with the steel plate. Secondly, the grooved steel plate was welded to the lower cover plate (Figure 3b), and then two additional grooved plates were welded to the previous plate to form a cross-shaped closed cavity (Figure 3c), and concrete was poured inside the cavity of the cross-shaped steel member. Finally, after 28 days of curing, the concrete on the upper surface of the specimen was polished and smoothed, and the upper cover plate and the hinge were welded to complete the production of the specimen (Figure 3d).

**Table 1.** Main parameters of specimens.

| No | Specimen | Axial Length $L$ (mm) | Slenderness Ratio $\lambda$ | Stiffener Type $s$ or $d$ (mm) | Steel Ratio $\alpha$ (%) | Load Eccentricity $e$ (mm) | Bearing Capacity $N_u$ (kN) | Ductility Coefficient [34] $\mu$ |
|---|---|---|---|---|---|---|---|---|
| 1 | CRa-140-0 | 1000 | 9.2 | 140 | 9.85 | 0 | 4600 | —— |
| 2 | CRa-140-27 | 1000 | 9.2 | 140 | 9.85 | 27 | 3544 | 6.85 |
| 3 | CRa-140-54 | 1000 | 9.2 | 140 | 9.85 | 54 | 2926 | 6.02 |
| 4 | CRb-140-0 | 1800 | 16.6 | 140 | 9.85 | 0 | 4602 | —— |
| 5 | CRb-140-27 | 1800 | 16.6 | 140 | 9.85 | 27 | 3504 | 4.84 |
| 6 | CRb-140-54 | 1800 | 16.6 | 140 | 9.85 | 54 | 2874 | 4.35 |
| 7 | CRb-200-0 | 1800 | 16.6 | 200 | 9.77 | 0 | 3753 | —— |
| 8 | CRb-200-27 | 1800 | 16.6 | 200 | 9.77 | 27 | 3483 | 4.88 |
| 9 | CRb-200-54 | 1800 | 16.6 | 200 | 9.77 | 54 | 2733 | 4.46 |
| 10 | CRc-140-0 | 2500 | 23.1 | 140 | 9.85 | 0 | 4483 | —— |
| 11 | CRc-140-27 | 2500 | 23.1 | 140 | 9.85 | 27 | 3473 | 4.10 |
| 12 | CRc-140-54 | 2500 | 23.1 | 140 | 9.85 | 54 | 2770 | 4.22 |
| 13 | Cb-27 | 1800 | 16.6 | No stiffening | 9.27 | 27 | 3628 | 2.67 |
| 14 | Cb-54 | 1800 | 16.6 | No stiffening | 9.27 | 54 | 2897 | 2.55 |
| 15 | CLb-50-27 | 1800 | 16.6 | 50 | 9.66 | 27 | 3698 | 2.79 |
| 16 | CLb-50-54 | 1800 | 16.6 | 50 | 9.66 | 54 | 2990 | 2.18 |
| 17 | CDb-27 | 1800 | 17.5 | Multi-cell form | 11.61 | 27 | 3986 | 5.49 |
| 18 | CDb-54 | 1800 | 17.5 | Multi-cell form | 11.61 | 54 | 3223 | 4.05 |

Note. $\lambda = L/i$, where $L$ and $i$ are the height of the specimen and the radius of gyration of the specimen, respectively [35]. $i = (E_sI_s + 0.2\ E_cI_c)^{1/2}/(E_s\ A_s + 0.2\ E_cA_c)^{1/2}$, where $I_s$ and $I_c$ are the moment of inertia of steel and concrete, respectively, and $E_s$ and $E_c$ are the modulus of elasticity of steel and concrete, respectively.

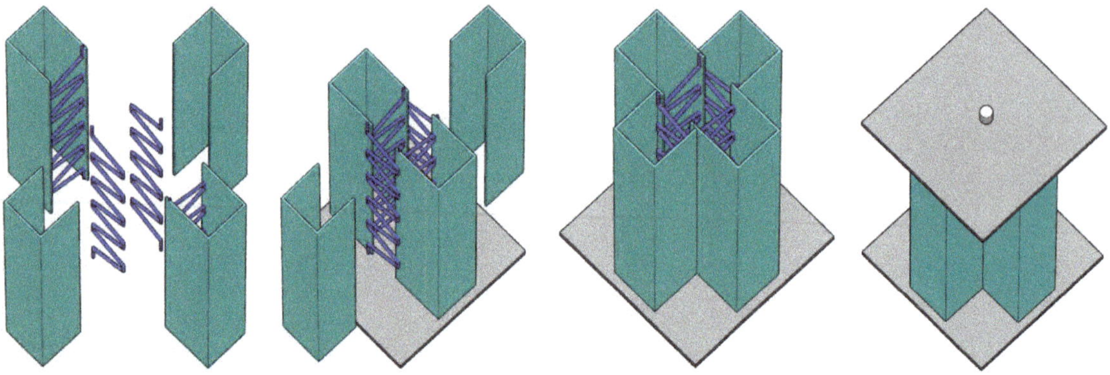
(**a**) Step 1 (**b**) Step 2 (**c**) Step 3 (**d**) Step 4

**Figure 3.** Processing schematic of truss-reinforced CCFST columns.

## *2.2. Material Properties*

The main mechanical property indexes of core concrete were determined in accordance with the Chinese standard GB/T 50081-2019 "Ministry of Housing and Urban-Rural Development of the People's Republic of China, Standard for Test Method of Concrete Physical and Mechanical Properties" [36]. The compressive strength $f_{cu}$ of the 150 mm × 150 mm × 150 mm cube test block was 44.1 MPa, the elastic modulus $E_c$ was 32.3 Gpa, and Poisson's ratio $\mu_c$ was 0.221. The tensile tests for the steel tube and steel bars were determined in accordance with the Chinese standard GB/T 228-2010 "Metallic Materials-Tensile Testing-Part 1: Method of Test at Room Temperature" [37]. Steel bars do not have an obvious yielding platform. Thus, the yield strength is determined by the 0.2%

proof stress. The measured mechanical properties of the steel are shown in Table 2, where $f_y$ and $f_u$ are the yield strength and tensile strength of the steel tube, respectively, and $E_s$ and $\mu_s$ are the elastic modulus and Poisson's ratio of the steel plates, respectively.

**Table 2.** Mechanical properties of steel.

| Type | $f_y$ (MPa) | $f_u$ (MPa) | $E_s$ (GPa) | $\mu_s$ |
|---|---|---|---|---|
| Steel plate ($t$ = 3.90 mm) | 304.2 | 444.3 | 208 | 0.281 |
| Steel bar ($\varphi$ = 7.24 mm) | 539.4 | 606.0 | 206 | 0.270 |

*2.3. Test Setup and Measurement*

Eccentric compression tests were carried out using a 30,000 kN universal testing machine (Figure 4). The machine came from Tianshui Hong Shan Testing Machine Co., Ltd., Tianshui, China, and the specification type was YAW-30000. During the tests, displacement control was applied at a constant rate of 0.4 mm/min. Upon reduction of the load to 70% of the peak load or when the deformation of the specimen was too large to continue loading, the test was stopped.

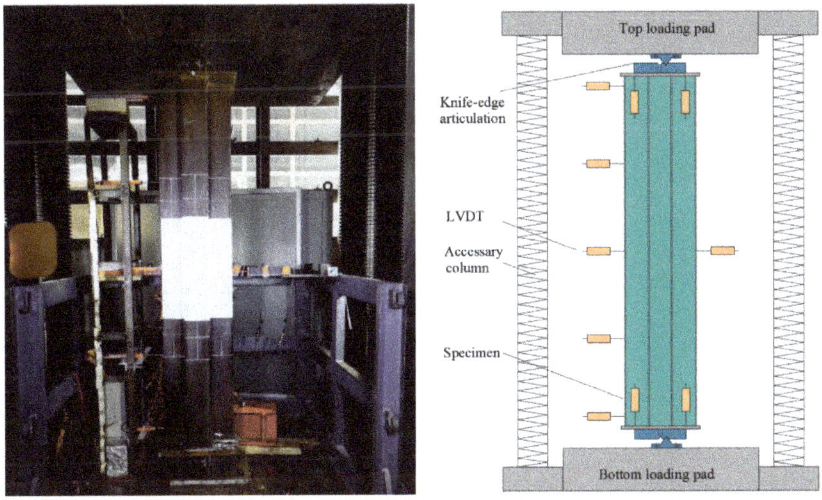


(a) Layout of loading device (b) Picture of test loading

**Figure 4.** Test setup.

The load data of the test were collected by the testing machine. The displacement data were taken by 12 linear variable displacement transducers (LVDTs) (Figure 5a). Specifically, axial deformation between two end caps was captured using 4 LVDTs, 5 LVDTs were installed in the height direction to gauge the horizontal deflection of specimens, 2 LVDTs were used to measure the horizontal displacement at concave corners of specimens, and, in addition, a No.10 LVDT was set up to measure the change at the mid-span of the specimen. The strains of steel plate were obtained by 23 groups of strain gauges of which 17 groups were arranged on the steel tube of the middle section of the specimen, and each group contained 2 strain gauges to measure horizontal and longitudinal strains. In addition, a total of 6 groups of strain gauges were set on the steel plates No.1 and No.2 at 45 mm above the middle section of the column. The specific layout is shown in Figure 5b,c.

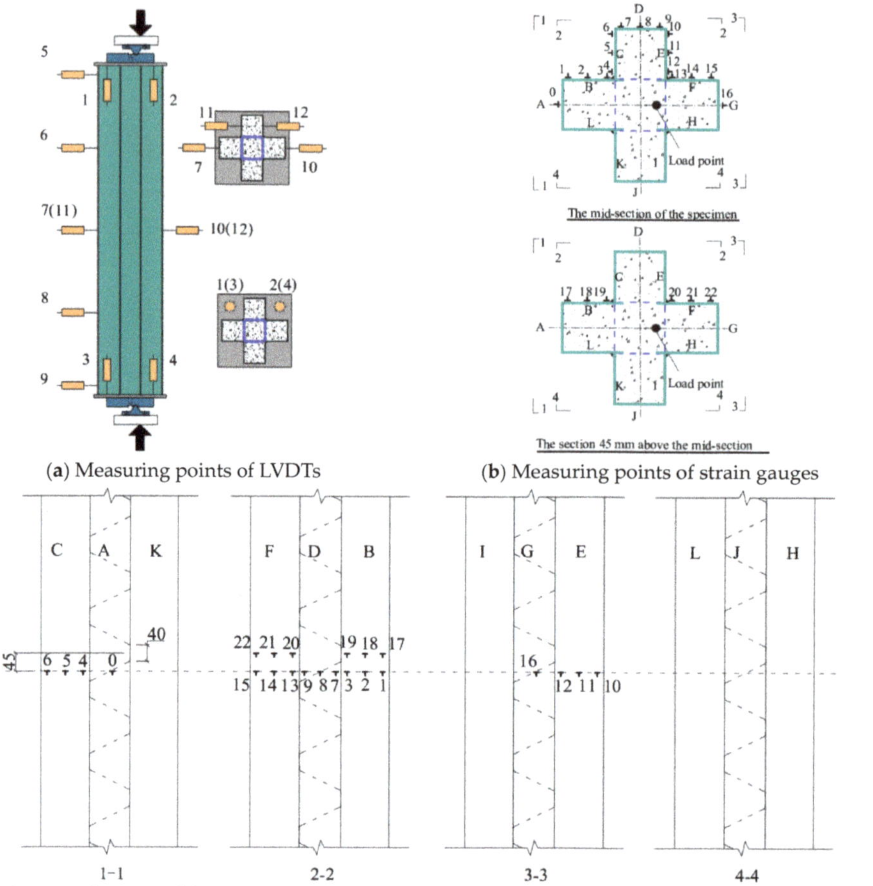


(**a**) Measuring points of LVDTs (**b**) Measuring points of strain gauges

(**c**) Schematic diagram of the positional relationship between the strain gauge and the reinforced truss

**Figure 5.** Arrangement of measuring points.

## 3. Test Results and Discussion

### 3.1. Failure Modes

Under eccentric compression, the CCFST specimens showed different failure modes with the change in slenderness ratio (Figure 6). For the short specimen CRa-140-27 with a height of 1000 mm, at the initial stage of loading, the deformation of the longitudinal and horizontal deformation were not significant. After gradually increasing the load to around 60% of the peak load, the specimen began to produce the "ZiZi" sound of the steel tube separated from concrete inside the specimen. When loaded to 95% of the peak load, the middle of the G-surface of the steel tube in the compression zone began to buckle slightly, while the overall horizontal deformation was not significant. As the load approached the peak value, slight buckling began to appear on the steel plates (F and H), the specimen showed a small bending deformation as a whole, and the horizontal deflection near the mid-section of the column gradually increased. After the peak load was reduced to 85%, the horizontal deformation was serious. There were obvious multi-wave bulging phenomena on the steel plates (G, F, E, and H) near the mid-section of the column, with the spacing of the wave peaks ranging from 120 mm to 150 mm. The short specimen CRa-140-27 showed the failure mode of the steel tube first by undergoing bulging, and then the overall bending deformation of the specimen occurred, which is a typical strength failure characteristic.

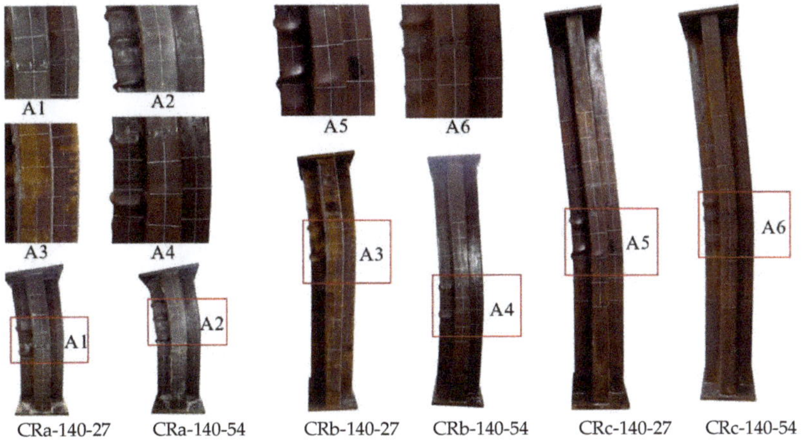


**Figure 6.** Failure phenomenon of steel tubes of truss-reinforced specimens.

For specimens with heights of 1800 mm and 2500 mm, the failure modes were generally close to each other (Figure 7). Taking the truss-reinforced specimen CRb-200-27 as an example, the eccentric compression test process was analyzed in detail. There was no discernible alteration in the appearance of the specimen from the initial stage of loading to 75% of a peak load, and the deflection growth was proportional to the load growth. With the continuous increase in load, the specimen began to show flexural deformation, and the change in horizontal deflection at the mid-section of column was more obvious than that of the column with a height of 1000 mm. As the peak load was approached, new plate buckling occurred 300 mm above the mid-section of the G-surface, and multiple small plate buckling occurred on the F-surface as well, with the bulge wave peaks spaced about 120 mm apart. As the displacement load increased further, the axial load gradually decreased, and the degree of bulging at the original steel plate bulge increased. When the load was reduced to 3300 kN (0.95 Nu), two new steel tube bulges were added on the E-surface, the peak of the bulge wave was staggered with the F-side, and the steel plate bulge also appeared on the D-surface at a point 70 mm above the mid-section of the column. The test was stopped when the load was reduced to 85% of the peak load, and then the truss-reinforced specimen showed a typical bending failure mode.

During the test of the unstiffened specimen Cb-27, a small local buckling first appeared in the middle of the G-surface. With the increase in vertical displacement load, significant steel tube bucklings were also observed (E, F, H, and I). At the end of the test, it was observed that the steel plates at concave corners near the mid-span of the column were pulled out by the expansion force of concrete. Following the cutting of steel tube (Figure 8), it was observed that there was an obvious crushing phenomenon on the plates (F and G); at the same time, lateral cracks appeared in the concrete in the middle of the A-surface. For truss-reinforced specimen CRb-200-27, the concrete on the F-surface, the G-surface had multiple crushings along the height of the column, and the crushing area corresponds to the buckling position of the steel plate. Multiple lateral cracks were observed in the concrete of the column limb in the tensile zone at the far end of the force. Compared to the unstiffened specimen Cb-27, the lateral cracks in the concrete of specimen CRb-200-27 had a wider distribution, the number of cracks increased significantly, and the cracks were more evenly distributed. Additionally, by observing specimens with multi-cell and spiral stirrup stiffening, it was noted that the multi-cell specimen showed multiple bulging waves within 200 mm above and below the mid-span of the column, resembling the form observed in the truss-reinforced specimen. The failure mode of specimens with spiral stirrup stiffening was similar to that of unstiffened specimens, characterized by the steel tube being extruded laterally due to the expansion of concrete and a severe lateral bulging of the compressed zone's steel tube. By comparing the phenomenon of lateral crack expansion in the tension section of the concrete on the A-surface during the failure of specimens with different configuration forms (Figure 8), it was evident that, for specimens with the same eccentricity ratio, those with reinforced truss exhibited a wider range of lateral crack expansion along the column height and smaller crack widths compared to unstiffened specimens.

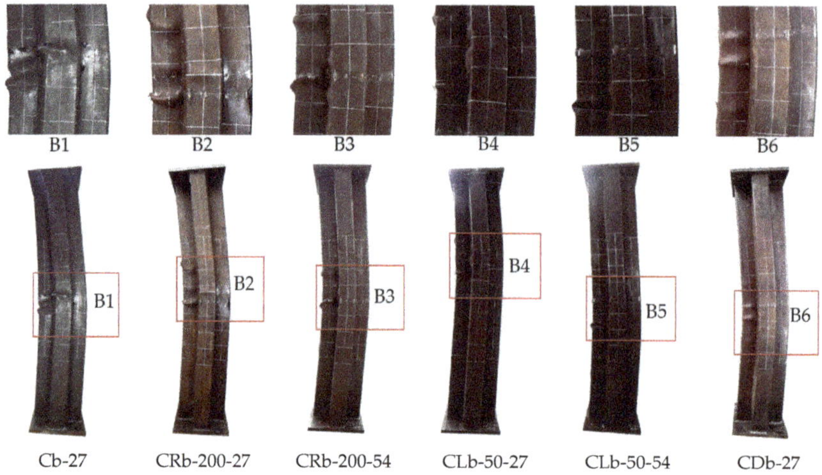


**Figure 7.** Failure phenomenon of specimens with different configuration forms.

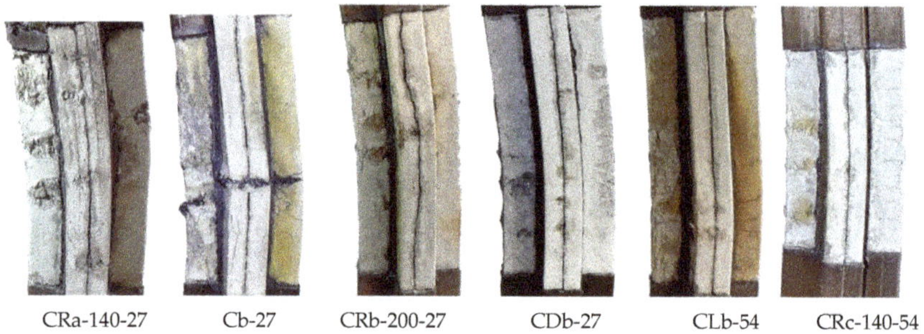


**Figure 8.** Failure phenomenon of concrete of specimens with different configuration forms.

The horizontal deflection distribution curves of the specimens reflect the characteristics of horizontal deformation. In this study, five horizontal displacement gauges were installed along the height direction of the specimen. Horizontal deflection curves under different load levels were plotted using deflection data (Figure 9). The vertical axis represents the height coordinates from the bottom of the column, and the horizontal axis represents horizontal deflection at different heights from the bottom of the column. $N_u$ is the ultimate load value of the specimen, and the "down" in brackets represents the descending segment. As shown in Figure 9, the bending deformation of all specimens generally follows a sine half-wave curve, and the development of the curves corresponds closely to the failure process of specimens. Before reaching the peak load, the horizontal deflection trends at the mid-span of all specimens were generally consistent. Upon gradual increase of the load to peak load, the differences in horizontal deflections became increasingly apparent. Compared with unstiffened specimen Cb-27, the horizontal deflection at the mid-section of specimen CRb-200-27 increased faster. When the load was reduced to 0.85 $N_u$, the deflection at the mid-span of specimen CRb-200-27 reached 15.3 mm, which was significantly higher than the 6.75 mm of the unstiffened specimen Cb-27. When the eccentricity ratio increased to 0.15, the horizontal deflection of truss-reinforced specimen CRb-200-54 was still growing faster than that of specimen Cb-54, and the deflection reached 20.7 mm and 12.1 mm, respectively. This shows that after the stiffened measure is adopted, the deformability of the member is enhanced, and the synergistic working of the steel tube and concrete is improved after the peak load.

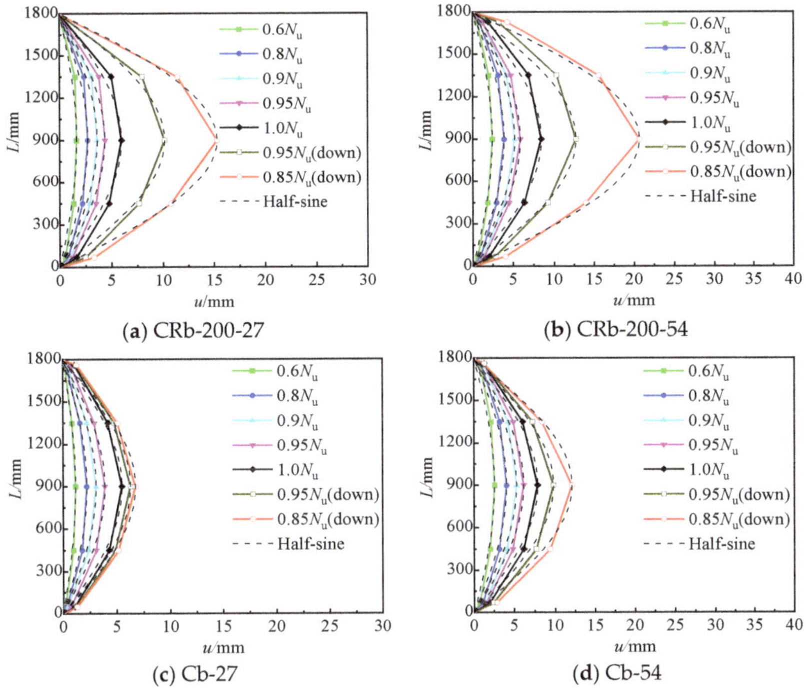


**Figure 9.** Deflection deformation curve of specimens along the height direction.

## *3.2. Influence of Parameters*

### 3.2.1. Configuration Forms

In this section, the load ($N$)–horizontal deflection ($u$) at 1/2 height relationship curves of different configuration forms (spiral stirrup stiffening, multi-cell, and reinforced truss stiffening) are plotted (Figure 10) to compare their flexural capacity and ductility performance. For the specimens with an eccentricity ratio of 0.08, in the elastic phase, the load ($N$)–horizontal deflection ($u$) at 1/2 height relationship curves exhibit linear development with increasing load, and the stiffness development of the three types of specimens is essentially consistent. Upon entering the elastic–plastic phase, the stiffness of the load ($N$)–horizontal deflection ($u$) at 1/2 height relationship curves for specimens CRb-200-27 (reinforced truss), CLb-50-27 (spiral stirrup stiffening), and CDb-27 (multi-cell) slowly decreases. At peak load, specimen CRb-200-27 has a lower bearing capacity, while specimens CLb-50-27 and CDb-27 exhibit increases of 6.2% and 14.4%, respectively, compared to specimen CRb-200-27. After reaching peak load, the load–deflection curve of specimen CLb-50-27 (spiral stirrup stiffening) rapidly descends, and it exhibits a ductility coefficient of 2.79, indicating poorer ductility. In contrast, specimens CRb-200-27 and CDb-27 achieve ductility coefficients of 4.88 and 5.49, respectively, showing improvements of 74.9% and 96.8% compared to specimen CLb-50-27, demonstrating better ductility. Spiral stirrup provides a certain amount of confinement to the internal concrete, which can increase the bearing capacity of the hooped concrete. But it has no obvious restriction effect on the steel at the concave corner, and then the steel tube and the concrete at the corner are separated, which is the primary reason for the worse ductility of specimens with spiral stirrup stiffening compared to those with reinforced truss and multi-cell forms.

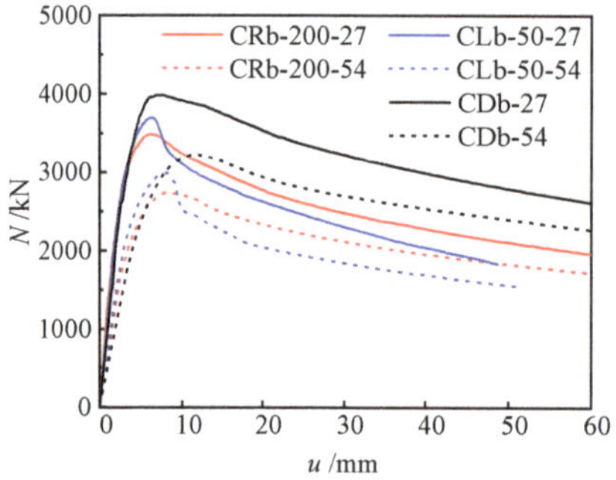


**Figure 10.** Load (*N*)–horizontal deflection (*u*) at 1/2 height relationship curves of different configuration forms.

When the eccentricity ratio increased to 0.15, the load (*N*)–horizontal deflection (*u*) at 1/2 height relationship curves of the steel-bar truss stiffening form, the spiral stirrup stiffening, and the multi-cell specimens were basically consistent with the above laws. The ductility coefficients of CRb-200-54 and CDb-54 were 4.46 and 4.05, respectively, while the ductility coefficient of CLb-50-54 was only 2.18, which indicated that the truss-reinforced specimen showed better ductility, even close to that of the specimen of the multi-cell form. In addition, the steel ratio of the reinforced truss form was about 9.8%, while that of the multicavity form was 11.6, and the former reduced the steel ratio by 1.8% in comparison with the latter, which saves steel.

### 3.2.2. Reinforced Truss Node Spacings

When the reinforced truss node spacing of the specimen tends to infinity, its mechanical properties approximate those of the unstiffened specimen. To investigate the influence of the reinforced truss node spacing on the mechanical properties of components, this section compares and analyzes the load (*N*)–horizontal deflection (u) at 1/2 height relationship curves of specimens with node spacings of infinity (unstiffened), 200 mm, and 140 mm (Figure 11). During the initial loading phase up to approximately 0.7 $N_u$, specimens with the same eccentricity exhibited similar stiffness development trends, with deflection growth proportional to the increase in load. As the load further increased towards the peak load, the specimens entered the elastoplastic stage, where the load (*N*)–horizontal deflection (*u*) at 1/2 height relationship curves exhibited nonlinear development, but the differences between the curves were not significant. When loaded to the peak load, the peak loads of the unstiffened specimen Cb-27, the specimen CRb-200-27 with a node spacing of 200 mm, and the specimen CRb-140-27 with a node spacing of 140 mm were 3628 kN, 3483 kN, and 3504 kN (Table 1), respectively, with a difference of no more than 5% between them. When the eccentricity increased to 54 mm, the peak load differences of these three types of specimens remained within 6%. Therefore, it can be inferred that changes in the reinforced truss node spacing had a minor impact on the eccentric compression bearing capacity of the specimens. However, once the load (*N*)–horizontal deflection (*u*) at 1/2 height relationship curves of the specimens entered the descending segment, the differences between different specimens became significant. Specifically, the curve of the unstiffened specimen experienced a steep drop, while the curves of the stiffened specimens with node spacings of 200 mm and 140 mm exhibited a gradual descending trend, demonstrating good deformability. Finally, the ductility coefficients of specimens Cb-27, CRb-200-27, and CRb-140-27 were calculated as 2.67, 4.88, and 4.84, respectively (Table 1). When the eccentricity increased to 54 mm, the ductility coefficients of these three types of specimens decreased to 2.55, 4.46, and 4.35, respectively. Therefore, the introduction of steel truss stiffening significantly affected the ductility of the specimens, with an improvement of over 70%. When the reinforced truss node spacing was reduced from 200 mm to 140 mm, the ductility improvement of the specimens was not significant, further indicating that a node spacing of 200 mm is sufficient to achieve significant enhancement in the ductility of the component.

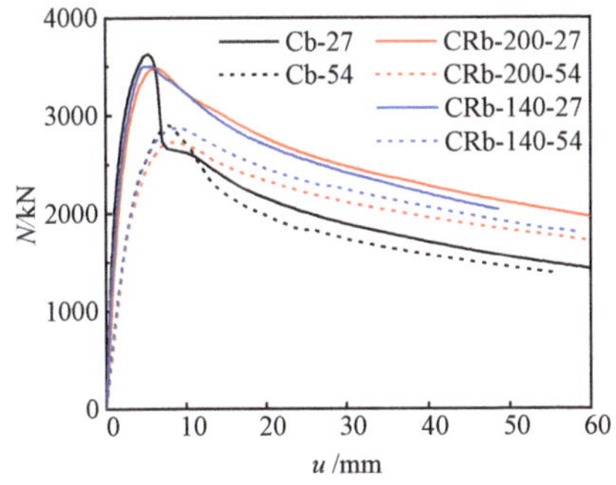


**Figure 11.** Load–deflection curves of typical specimens.

### 3.2.3. Slenderness Ratios and Eccentricity Ratios

To investigate the influence of the slenderness ratio on the eccentric compression behavior of truss-reinforced specimens, a comparative analysis was conducted on the load ($N$)–horizontal deflection ($u$) at 1/2 height relationship curves of specimens with different slenderness ratios ($\lambda$ = 9.2, 16.6, 23.1) (Figure 12a). In the initial loading stage, specimens with larger slenderness ratios exhibited lower initial stiffness. While the load increased to the peak load, specimens with larger slenderness ratios showed greater horizontal deflection at the mid-span and smaller peak load. Once the specimen curves entered the descending segment, the curves of specimens with different slenderness ratios all declined relatively smoothly, indicating the good ductility of the truss-reinforced specimen. Specifically, the ductility coefficients of specimens CRa-140-54 ($\lambda$ = 9.2), CRb-140-54 ($\lambda$ = 16.6), and CRc-140-54 ($\lambda$ = 23.1) reached 6.02, 4.35, and 4.22, respectively. Additionally, a comparative analysis was also conducted on the load ($N$)–horizontal deflection ($u$) at 1/2 height relationship curves of specimens with different eccentricities ($e$ = 27 mm and 54 mm) (Figure 12b). It was found that with the increase in eccentricity, the elastic segment became shorter, the bearing capacity decreased, and the curve declined more smoothly after reaching the peak load.

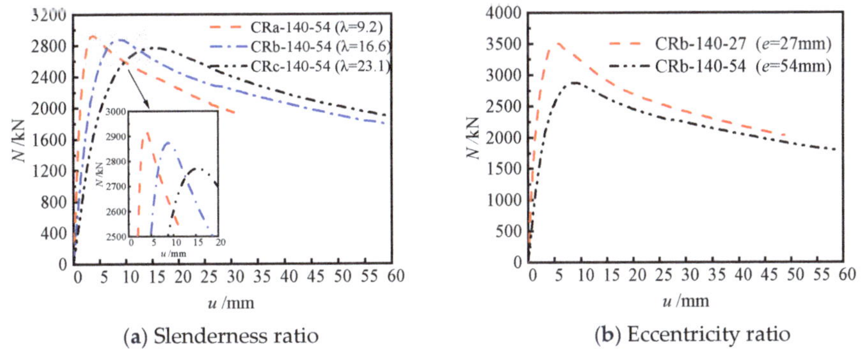


**Figure 12.** Load ($N$)–horizontal deflection ($u$) at 1/2 height relationship curves of specimens with different slenderness ratios and eccentricity ratios.

## 3.3. Enhancing Mechanism of Steel-Bar Truss

### 3.3.1. Stress of Steel Tube at Concave Corners

To further elucidate the restraining effect of the reinforced truss on the steel tube at the concave corner and the confinement mechanism of the steel tube on concrete for the CCFST column, the strains of the steel plates at concave corners of unstiffened and stiffened specimens were monitored. The measuring points of the unstiffened specimens were arranged similarly to the strain gauges at the mid-span of the truss-reinforced specimen (Figure 13a). Additionally, based on the setting of the reinforced truss along the height, strain monitoring was conducted at another critical section (Figure 13b). Ultimately, the strain data were converted into stress using an elastoplastic analysis method [38], yielding the stress curves of the steel tubes under eccentric compression. $\sigma_v$ and $\sigma_h$ are the vertical and horizontal stresses of the steel plate, respectively. $\sigma_z$ is not the stress of the steel plate, but rather the equivalent stress obtained from the calculation of vertical and horizontal stresses, $\sigma_z = \sqrt{\sigma_h^2 + \sigma_v^2 - \sigma_h\sigma_v}$.

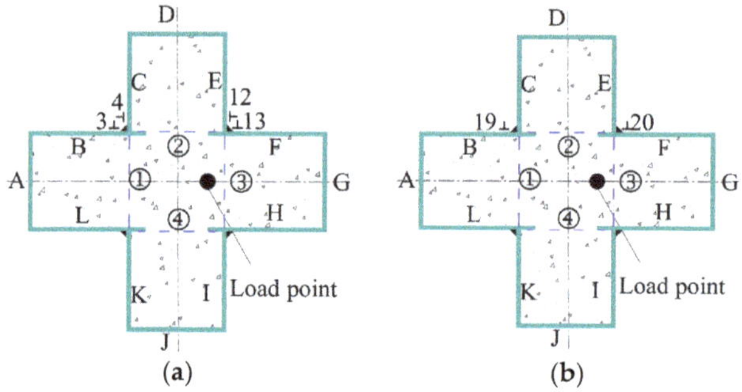


**Figure 13.** Schematic diagram of the measuring points at the concave corner of the truss-reinforced specimen. (**a**) The measuring points at the mid-section of the specimen. (**b**) The measuring points 45 mm away from the mid-section along the height of the specimen.

For unstiffened specimen Cb-54 (Figure 14), before the peak load, steel plates at the concave corner were mainly under longitudinal compressive stresses, the horizontal stress was small, and the synergistic action between the steel tube and the concrete at the concave corner was not obvious. After the peak load, the horizontal stress development at each measuring point was still slow, and when the specimen was destroyed, the horizontal tensile stress at points 3#, 4#, and 12# was 1.9 MPa, 22.6 MPa, and 66.2 MPa, respectively, which was in line with the general rule that the horizontal stress of steel plate had a small increase after the extrusion of the side wall of the steel tube by concrete expansion under compressive loads. As for measuring point 13#, it was subjected to horizontal and longitudinal two-way compressive stress, and the confinement effect on concrete was weaker. This shows that the steel plate of the unstiffened specimen can only resist the horizontal deformation of concrete through its own stiffness, and lacks effective horizontal confinement to concrete, which cannot effectively exert the combined effect of the steel tube and concrete.

For specimen CRb-200-54 ($s$ = 200 mm, $e$ = 54 mm) (Figure 15), before the peak load, the development of horizontal stresses at concave corners of the steel tube was not obvious, and the effect of the steel bars was not demonstrated. At peak loads, the horizontal stresses gradually began to develop. When the load decreased to 2713 kN (0.99 $N_u$), the horizontal tensile stress at point 13# increased rapidly to 200.0 MPa; meanwhile, the horizontal tensile stress at point 4# also increased to 98.1 MPa when the load decreased to 2322 kN (0.85 $N_u$). This shows that the addition of diagonal bars can effectively restrict the buckling of the steel tube at concave corners, enhance the confinement of concrete by the steel tube, and realize the synergistic working of the steel tube and concrete.

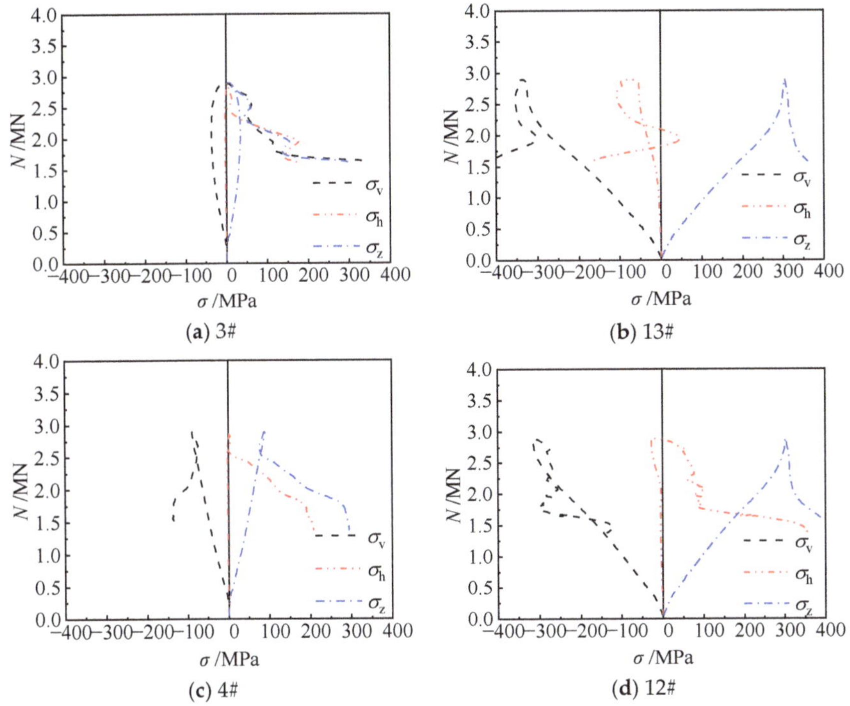


**Figure 14.** Stress of steel plate at the concave corner of specimen Cb-54.

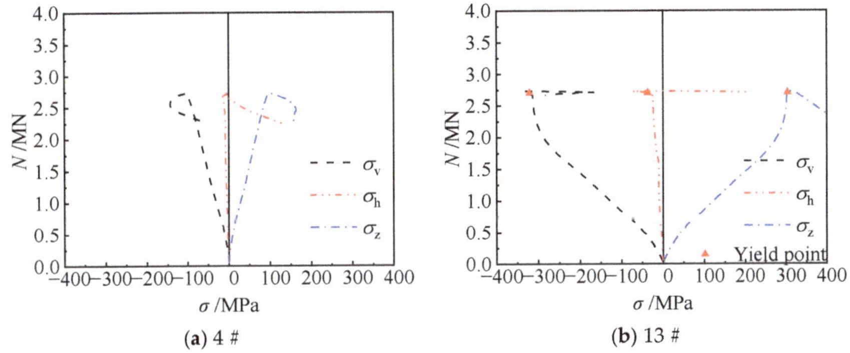


**Figure 15.** Stress of steel plate at the concave corner of specimen CRb-200-54.

In order to further illustrate the development of the stresses in the steel tube at concave corners along the column height direction, a section 45 mm away from the mid-section of the specimen was selected as the critical section. By observing the horizontal stress curves of steel plates of specimen CRb-200-27 and specimen CRb-140-54 (Figure 16), it was found that before reaching the peak load, the horizontal stress at points 19# and 20# developed slowly. During the process from peak load to specimen failure, the horizontal stress at point 19# developed rapidly, with the maximum horizontal tensile stress reaching 356 MPa and 176 MPa for specimens CRb-200-27 and CRb-140-54, respectively. The tensile stress at point 20# gradually appeared only after the specimen was damaged. The primary reason is that at point 19#, the steel plate was directly subjected to the in-plane tensile force generated by the diagonal bar ②. As for point 20#, the point was located in the position between the two adjacent welding points of the diagonal bar ②, and the diagonal bar did not act directly on the steel plate at this point, so that the influence of the steel bars on the internal tension in the steel plate was small, and was only visible when the specimen underwent significant damage and deformation.

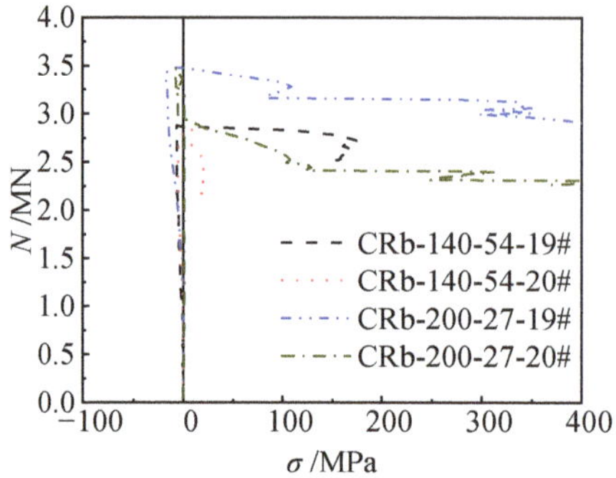


**Figure 16.** Horizontal stress profile at the concave corner 45 mm above the mid-section.

When the slenderness ratio of the specimens was reduced to 9.2 (Figure 17), for the truss-reinforced specimen CRa-140-27 ($\lambda$ = 9.2, $e$ = 27 mm), prior to peak load, the horizontal tensile stress at point 13# developed slowly; the horizontal tensile stress began to develop rapidly and had grown to 332 MPa when the load decreased to 0.94 $N_u$. When the eccentricity ratio increased to 0.15, the horizontal tensile stresses at point 13# of specimen CRa-140-54 ($\lambda$ = 9.2, $e$ = 54 mm) increased gradually. When the specimen was destroyed, the horizontal tensile stress at point 13# reached 168 MPa.

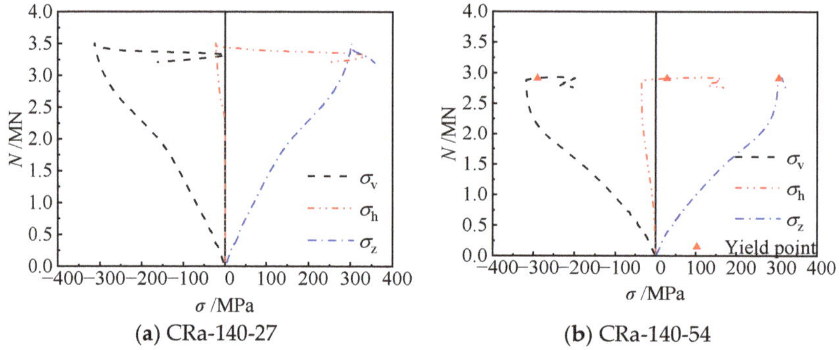


**Figure 17.** Stress of point 13# at the concave corner of specimens.

In summary, the reinforced truss has a certain restraining effect on the steel tubes at concave corners of the member. Steel-bar trusses arranged along the bending direction of the member can effectively enhance the confinement of concrete by the steel tube, thereby improving the overall eccentric compression behavior of the member.

### 3.3.2. Longitudinal Strain Distribution of Steel Tube

The damage of the specimens mainly occurred near the column 1/2 height and the bending moment of the specimens and the longitudinal strain values of the steel tubes were the largest here; therefore, the longitudinal strains at this height are discussed. The longitudinal strain load distributions of steel tubes for typical specimens Cb-54 and CRb-200-54 are shown in Figures 18 and 19, where positive strains are tensile strains and negative strains are compressive strains.

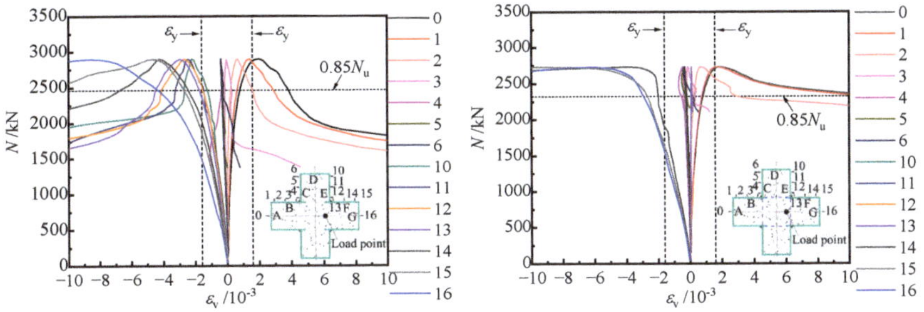


(**a**) Cb-54 (**b**) CRb-200-54

**Figure 18.** Longitudinal strain curve of steel tube with mid-section.

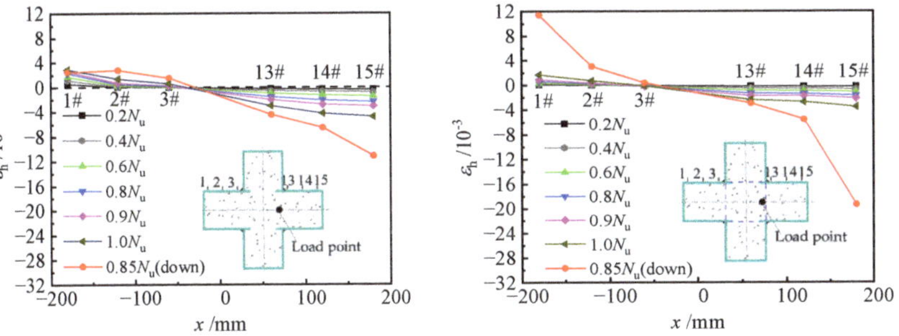


(**a**) Cb-54 (**b**) CRb-200-54

**Figure 19.** Longitudinal strain distribution at the mid-span.

Figure 18 gives the longitudinal strain relationship curves for the steel tube at the mid-section of the specimen, with a yield strain $\varepsilon_y$ of 1590 microstrains. The curve development trend corresponds to the specimen's failure process. In the early stages of loading, the steel tube strain increasedlinearly with increasing load. As loading continued, both the positive and negative bending of the specimen entered the plastic stage starting from the specimen's edge. Firstly, the longitudinal strain of the G-surface at the edge of the compression zone began to increase rapidly, and reached the yield strain first, and then the strain at points 15~10 also reached the yield strain in turn (Figure 18). When the load reached a peak, the tensile steel plate A-surface at the distal end of the two typical specimens Cb-54 and CRb-200-54 also reached the yield strain almost at the same time; the longitudinal stresses in the column mid-section of specimens Cb-54 and CRb-200-54 varied differently upon peak load. The longitudinal strain curves of truss-reinforced specimen CRb-200-54 dropped more gently as compared with that of Cb-54, with a larger increase in longitudinal strains in the B-surface of the tensile plate and the F-surface of the compressive steel plate when the load of the specimen decreased to 0.85 $N_u$. The longitudinal strain curves of the compression plate F-surface decreased more gently, and the longitudinal strain of the tensile A-surface had increased greatly when the load of the specimen decreased to 0.85 $N_u$. It can also be observed more obviously through Figure 19 that the longitudinal strain distribution patterns of typical specimens were basically the same in different load stages before a peak load and were in line with the plane section assumption. Only after a peak load did the longitudinal strains in the tensile and compressive zones of truss-reinforced specimen CRb-200-54 have a large increase, and when the specimen was loaded up to 0.85 $N_u$ (down), the steel plate's No.0 tensile zone and No.16 compressive zone reached 11,427 and 19,308 microstrains, respectively, which indicates that the configuration of reinforced truss can better improve the flexural deformability of the member.

### 3.3.3. Horizontal Strain Distribution

Figure 20 illustrates the distribution of horizontal strains along the cross-sectional height of typical specimens under different load levels. It can be observed that before loading to 0.6 $N_u$, the horizontal strains along the cross-section height of the specimen exhibited an approximately linear distribution. As the load increased to the peak load, the horizontal tensile strains at measuring points 13#, 14#, and 15# increased gradually, the horizontal strains at measuring points 1#, 2#, and 3# changed less, while the horizontal tensile strain at measuring points 13#, 14#, and 15# increased to some extent. Compared with unstiffened specimen Cb-27, the horizontal strain of truss-reinforced specimen CRb200-27 increased more during loading to 0.8 $N_u$ to 1.0 $N_u$. When loaded to the peak load, the horizontal tensile strains continued to increase in the compression zone at the measuring points 13#, 14#, and 15#, and the increase in specimen CRb200-27 was particularly significant. The main reason for the above law is that under eccentric compression, the longitudinal strain on the B-surface is smaller compared to that on the F-surface, and the horizontal deformation of the concrete in the column limb of the compression zone on the side of the F-surface is larger and squeezes the steel plate in two directions perpendicular to F-surface and G-surface, which leads to an increase in the horizontal strain at points 13#, 14#, and 15#. Meanwhile, the out-of-face bulging deformation occurred at the F-surface, which caused the horizontal strain at point 14# to be larger compared with points 13# and 15#. Specimen CRb200-27 had a diagonal bar ① welded near measuring point 13#. The diagonal bar ① acted on the F-surface of the steel plate to prevent the F-surface from being pulled out in the direction parallel to the F-surface. As a result, compared to the unstiffened specimen Cb-27,the addition of the diagonal bar ① increased the horizontal strain of the steel plate F-surface. Especially post peak load, the deformation of concrete was further intensified and then squeezed the steel plate G-surface and F-surface, whereas the steel bar ② limited the out-of-face displacement of the steel plate F-surface at the concave corner. Accordingly, the confinement effect of steel tube on concrete of the column limb was enhanced, especially after the peak load.

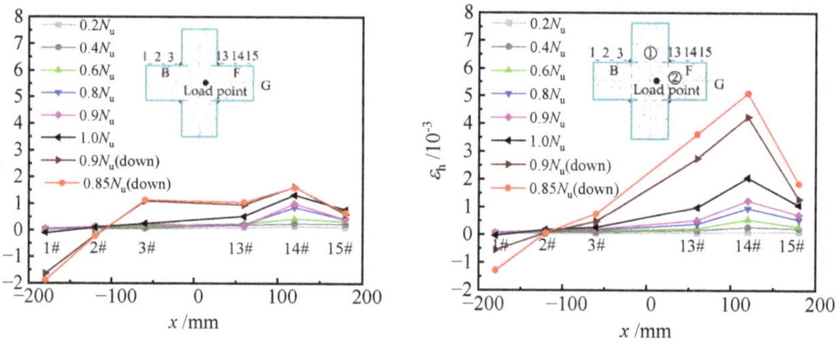


(**a**) Cb-27 (**b**) CRb200-27

**Figure 20.** Horizontal strain distribution at the mid-span.

## 4. Comparison with Existing Design Codes

Currently, there are various design codes to calculate the bearing capacity of composite members under combined compression and bending in China and abroad, and there is a large difference between calculation methods. Domestically, there are GB50936 [39], CECS159 [40], etc., and foreign design codes on CFST include Eurocode-4 [41], AISC360 [42], and so on. This paper focuses on the relevant calculations in the four design codes mentioned above.

The solution to GB50936 consists of two terms relating to axial forces and moments.

$$-\frac{N}{2.17N_u}+\frac{\beta_m M}{M_u(1-0.4N/N_E')}\leq 1 \quad \frac{N}{N_u}<0.255 \tag{1}$$

$$\frac{N}{N_u}+\frac{\beta_m M}{1.5M_u(1-0.4N/N_E')}\leq 1 \quad \frac{N}{N_u}\geq 0.255 \tag{2}$$

where $N_u$ is the stable load of composite columns under axial force; $M_u$ is the load-carrying capacity in bending; $\beta_m$ is the equivalent bending moment coefficient; and $N'_E$ is the Euler load of the column considering the partial coefficient of the material.

The solution to CECS159 is provided by

$$\frac{N}{\varphi_x N_u}+(1-\alpha_c)\frac{\beta_m M_x}{(1-0.8\frac{N}{N'_{Ex}})M_u}\leq 1 \tag{3}$$

$$\frac{\beta_m M_x}{(1-0.8\frac{N}{N'_{EX}})M_u}\leq 1 \tag{4}$$

$$\frac{N}{\varphi_y N_0}+\frac{\beta_m M_x}{1.4M_u}\leq \frac{1}{\gamma_0} \tag{5}$$

where $\gamma_0$ is the configuration importance coefficient, 1.0 in the case of this study; $\varphi_x$ and $\varphi_y$ are the stability coefficients under the axial compression in the plane of bending moment action and out of the plane of bending moment action, respectively; $N_0$ is the load-carrying capacity in compression; $\alpha_c$ is the percentage of load-carrying capacity shared by concrete; and $N'_{EX}$ is the Euler load of the column considering the partial coefficient of the material.

The *N-M* curve (Figure 21) in Eurocode-4 can be simplified into three straight lines for the design value of the plastic compression capacity of the section and for the design value of the plastic bending capacity of the section. $M_{maxRd}$ is the maximum design plastic resistance moment, $N_{pmRd}$ is the design axial force of the concrete section, and $M_{Ed}$ and $N_{Ed}$ are the axial bending moment and the axial force, respectively.

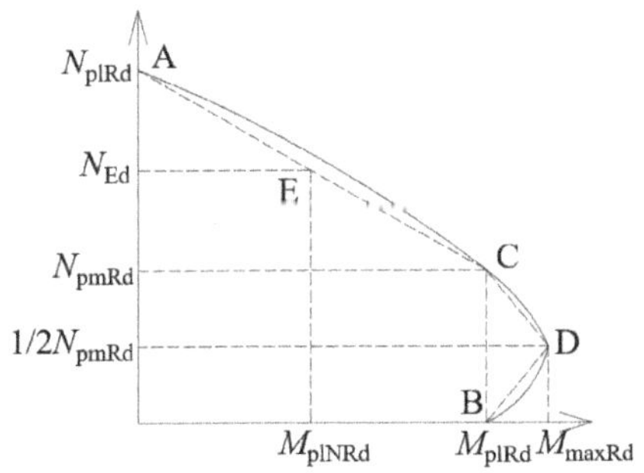


**Figure 21.** *N-M* interaction curve.

Composite members should satisfy the following conditions under compression–bending combination:

$$\frac{M_{Ed}}{M_{plNRd}}=\frac{M_{Ed}}{\mu_d M_{plRd}}\leq \alpha_m \tag{6}$$

$$\frac{N_{Ed}}{\chi N_{plRd}}\leq 1 \tag{7}$$

where $M_{plRd}$ and $M_{plNRd}$ are the plastic resistance moments of the cross-section without considering the axial force and considering the axial force, respectively; $M_{Ed}$ is the greatest moment in the member; $\alpha_m$ is a coefficient related to the strength grade of the steel, which is taken to be 0.9 for steel grades between Q235 and Q345; and $\chi$ is a reduction factor related to the slenderness ratio.

According to AISC360, composite sections are divided into noncompact, compact, and slender based on the width-to-thickness ratio of the steel section. The analysis methods for these different types of composite sections vary slightly. In this study, all the cross-sections considered are classified as compact sections. The solution for compact sections is then given by:

$$\frac{N}{N_u}+\frac{8}{9}\frac{M}{M_u}\leq 1 \quad \frac{N}{N_u}\geq 0.2 \tag{8}$$

$$\frac{N}{2N_u}+\frac{M}{M_u}\leq 1 \tag{9}$$

Code AISC360 is similar to CECS159, which does not consider the confinement effect of concrete by the steel tube.

In this paper, the four calculation methods mentioned in the previous section were used to obtain the eccentric compression bearing capacity of the members, and the calculations were compared with the test results (Table 3), so as to compare the accuracy of the calculation methods in each code. It should be noted that the relevant parameters in the code use the design value, while this article uses the standard value. In addition, due to the initial defects of specimen CRb-200-0 or an error during the test, bending deformation occurred during the loading process, which led to the low bearing capacity of the test. Therefore, when comparing with the existing design specifications, the data of the specimen were not considered.

**Table 3.** Comparison of test results with calculations from relevant codes.

| Specimen | $N_{tesk}$ /kN | $N_{EC}$ /kN | $N_{AISC}$ /kN | $N_{GB}$ /kN | $N_{CECS}$ /kN | $\frac{N_{EC}}{N_{test}}$ | $\frac{N_{AISC}}{N_{test}}$ | $\frac{N_{GB}}{N_{test}}$ | $\frac{N_{CECS}}{N_{test}}$ |
|---|---|---|---|---|---|---|---|---|---|
| CRa-140-0 | 4600 | 4415 | 3989 | 4893 | 4284 | 0.960 | 0.867 | 1.064 | 0.931 |
| CRa-140-27 | 3544 | 3608 | 2500 | 3546 | 3322 | 1.018 | 0.705 | 1.001 | 0.937 |
| CRa-140-54 | 2926 | 3050 | 1820 | 2780 | 2714 | 1.042 | 0.622 | 0.950 | 0.928 |
| CRb-140-0 | 4602 | 4399 | 3941 | 4870 | 4220 | 0.956 | 0.856 | 1.058 | 0.917 |
| CRb-140-27 | 3504 | 3595 | 2481 | 3529 | 3273 | 1.026 | 0.708 | 1.007 | 0.934 |
| CRb-140-54 | 2874 | 3040 | 1809 | 2768 | 2677 | 1.058 | 0.629 | 0.963 | 0.931 |
| CRb-200-27 | 3483 | 3595 | 2481 | 3529 | 3273 | 1.032 | 0.712 | 1.013 | 0.940 |
| CRb-200-54 | 2733 | 3040 | 1809 | 2768 | 2677 | 1.112 | 0.662 | 1.013 | 0.980 |
| CRc-140-0 | 4483 | 4315 | 3876 | 4836 | 4134 | 0.963 | 0.865 | 1.079 | 0.922 |
| CRc-140-27 | 3473 | 3527 | 2455 | 3517 | 3208 | 1.016 | 0.707 | 1.013 | 0.924 |
| CRc-140-54 | 2770 | 2982 | 1796 | 2752 | 2628 | 1.077 | 0.648 | 0.994 | 0.949 |
| Mean value | | | | | | 1.024 | 0.726 | 1.014 | 0.936 |
| Coefficient of variation | | | | | | 0.049 | 0.129 | 0.039 | 0.018 |

With the increase in the eccentricity, there is a tendency to increase the ratio of calculated to experimental values of EC4, and in contrast, the ratio of the calculated value of GB50936 to the test value has a tendency to decrease. By comparing the calculation results of CECS159, it can be found that the calculated values are smaller than the test values, which are on the safe side, with an mean deviation within 7% and a small dispersion.

In summary, the results obtained using EC4, GB50936, and CECS 159 were in perfect correspondence with the test values. The mean values of the three methods were 1.024, 1.014, and 0.936 and the coefficients of variation were 0.049, 0.039, and 0.018, respectively, which indicate high stability and accuracy in predicting the bearing capacity. Therefore, within the parameter range of this paper, it is recommended to adopt the calculation methods in EC4, GB50936, and CECS 159 to calculate the eccentric compression bearing capacity of truss-reinforced CCFST columns.

## 5. Conclusions

This paper presents a detailed investigation into the eccentric compression tests on the CCFST columns. The experimental procedures and observations are thoroughly described, focusing on the eccentric compression behavior of truss-reinforced columns. The obtained results include the load ($N$) versus horizontal deflection ($u$) curves, load–stress curves, and load–strain curves for the steel tube. Comparative analyses are conducted for various test parameters including the reinforced truss node spacing, slenderness ratio, and eccentricity ratio. Based on the study, the following conclusions are drawn:

(1) The reinforced truss is effective in enhancing concrete confinement by the steel tube, leading to significantly improved eccentric compression behavior for the CCFST. The failure mode of the steel tube experienced multiple-wave buckling rather than single-wave buckling for the truss-reinforced CCFST columns, and the concrete experienced multi-section crushing rather than single-section crushing.

(2) The reinforced truss could significantly improve the deformation ability of the eccentrically compressed CCFST column, and the ductility coefficient was increased by 70~80% compared with the unstiffened column, which could be close to that of the multi-cell member. In addition, the ductility was not changed much when the reinforced truss node spacing was reduced from 200 mm to 140 mm, and it is concluded that the ductility requirement for the member could be satisfied when the node spacing is 200 mm.

(3) The addition of the reinforced truss can restrict the out-of-plane displacement at concave corners of the steel tube, enhance the confinement of the steel tube to the core concrete, improve the overall eccentric compression behavior of the member, and enhance the synergistic working of the two materials, steel tube and concrete.

(4) By comparing the design code methods and test results, it is concluded that results obtained using EC4, GB50936, and CECS 159 are in perfect correspondence with the test values. Therefore, within the range of parameters in this paper, it is recommended to adopt the calculation methods in Chinese codes GB50936 and CECS159, and in Eurocode 4 to calculate the eccentric compression bearing capacity of truss-reinforced CCFST columns.

**Author Contributions:** Conceptualization, Y.T., S.Z., C.G. and Z.H.; methodology, Y.T., C.G., S.Z. and X.L.; software, Y.T. and G.X.; validation, Y.T., C.G. and S.Z.; formal analysis, Y.T., S.Z. and X.L.; investigation, Y.T.; resources, Y.T.; data curation, Y.T.; writing—original draft preparation, Y.T.; writing—review and editing, C.G., S.Z. and X.L.; visualization, Y.T.; supervision, S.Z.; project administration, Y.T. and X.L.; funding acquisition, Y.T., S.Z. and C.G. All authors have read and agreed to the published version of the manuscript.

**Funding:** This study is funded by the Special Projects of Science and Technology of CHINA MINMETALS CORPORATION (2019ZXB07).

**Institutional Review Board Statement:** Ethical approval is not required.

**Informed Consent Statement:** Not applicable.

**Data Availability Statement:** All data included in this study are available from the corresponding author upon request.

**Conflicts of Interest:** Authors Gaopeng Xiong, Chao Gong, Zhaoxin Hou are employed by the company Central Research Institute of Building and Construction Co., Ltd., MCC Group, Shenzhen 518055, China. The remaining authors declare that the research was conducted in the absence of any commercial or financial relationships that could be construed as a potential conflict of interest.

## References

1. Lei, M.; Shen, Z.; Li, Y.; Luo, J. State-of-the-art of concrete-filled specially shaped steel tubes. *Struct. Eng.* **2013**, *29*, 155–163. (In Chinese)
2. Zheng, Y.; Zeng, S. Flexural behaviour of stiffened and multi-cell L-shaped CFSTs considering different loading angles. *J. Constr. Steel Res.* **2021**, *178*, 106520. [CrossRef]

3. Zheng, Y.; Zeng, S. Design of L-shaped and T-shaped concrete-filled steel tubular stub columns under axial compression. *Eng. Struct.* **2020**, *207*, 110262. [CrossRef]
4. Ren, Q.-X.; Han, L.-H.; Lam, D.; Hou, C. Experiments on special-shaped CFST stub columns under axial compression. *J. Constr. Steel Res.* **2014**, *98*, 123–133. [CrossRef]
5. Han, L.-H.; Hou, C.-C.; Xu, W. Seismic performance of concrete-encased column base for hexagonal concrete-filled steel tube: Numerical study. *J. Constr. Steel Res.* **2018**, *149*, 225–238. [CrossRef]
6. Cui, W.; Wang, L.; Chen, H.; Chen, H.; Liu, S. The analysis on the seismic behavior of T-shaped prefabricated special-shaped CFST with rectangular multi-cell columns. *Structures* **2020**, *28*, 803–815. [CrossRef]
7. Hassam, M.; Guo, L.; Wang, Y. Experimental and numerical investigation of cross-shaped stub CFSTs under axial compression. *Mag. Concr. Res.* **2021**, *73*, 1225–1240. [CrossRef]
8. Sina, K.; Dongxu, L.; Brian, U. Compact and slender box concrete-filled stainless steel tubes under compression, bending, and combined loading. *J. Constr. Steel Res.* **2021**, *184*, 106813.
9. Lee, H.-J.; Choi, I.-R.; Park, H.-G. Eccentric Compression Strength of Rectangular Concrete-Filled Tubular Columns Using High-Strength Steel Thin Plates. *J. Struct. Eng.* **2017**, *143*, 04016228. [CrossRef]
10. Kristýna, H.; Petr, C.; Sabina, O.; Dalibor, K. Analysis of concrete taken from a loaded column by determining the modulus of elasticity. *Acta Polytech. CTU Proc.* **2024**, *47*, 47–52.
11. Přemysl, P.; Petr, M.; Petr, L. Experimental study of high thin-walled cold-rolled Z cross-sections purlins. *J. Constr. Steel Res.* **2023**, *208*, 108017.
12. Tu, Y.-Q.; Shen, Y.-F.; Zeng, Y.-G.; Ma, L.-Y. Hysteretic behavior of multi-cell T-Shaped concrete-filled steel tubular columns. *Thin-Walled Struct.* **2014**, *85*, 106–116. [CrossRef]
13. Cheng, R.; Hu, C.; Gong, M.; Wang, Y. Behaviors of improved multi-cell T-shaped concrete-filled steel tubular columns under eccentric loads. *J. Constr. Steel Res.* **2022**, *193*, 107251. [CrossRef]
14. Zhang, N.; Shen, Y.F.; Li, P.; Tu, Y.Q. Study on performance of multi-cell composite T-shaped concrete-filled steel tubular slender columns under axial compression. *J. Build. Struct.* **2015**, *36*, 254–261. (In Chinese)
15. Liu, L.L.; Tu, Y.Q.; Ye, Y.H. Experimental study of the properties of axially loaded multi-cell T-shaped concrete-filled steel tubular stub columns. *China Civ. Eng. J.* **2011**, *44*, 9–16. (In Chinese)
16. Huang, H.; Luo, S.Y.; Yang, C.; Xie, L.; Xu, C.G. Experimental study on mechanical behavior of combined cruciform concrete filled steel tubular column under eccentric compression. *J. Exp. Mech.* **2020**, *35*, 309–318.
17. Du, G.F.; Xu, L.H.; Xu, H.R.; Wen, F. Study on composite T-shaped concrete filled steel tubular columns under eccentric compression. *J. Build. Struct.* **2010**, *31*, 72–77. (In Chinese)
18. Sui, Y.; Tu, Y.; Guo, Q.; Zhang, J.; Ke, F. Study on the behavior of multi-cell composite T-shaped concrete-filled steel tubular columns subjected to compression under biaxial eccentricity. *J. Constr. Steel Res.* **2019**, *159*, 215–230. [CrossRef]
19. Li, Q.; Zhou, X.J.; Li, G.Q.; Liu, Z.; Wang, Z.; Wang, X.-B.; Xian, G.-D. Experimental study on the behavior of special T-shaped composite columns with concrete-filled square steel tubulars under eccentric loads. *J. Civ. Environ. Eng.* **2021**, *43*, 102–111. (In Chinese) [CrossRef]
20. Lin, Z.; Shen, Z.; Luo, J.; Zhang, J. Study on behavior of L-shaped concrete-filled steel tube stubs subjected to axial-compression. *Prog. Steel Build. Struct.* **2009**, *11*, 14–19. (In Chinese)
21. Zheng, Y.; Lai, P. Experimental behavior of T-shaped concrete-filled steel tubular columns under diagonal cyclic loading. *J. Constr. Steel Res.* **2020**, *169*, 106037. [CrossRef]
22. Liu, X.; Xu, C.; Liu, J.; Yang, Y. Research on special-shaped concrete-filled steel tubular columns under axial compression. *J. Constr. Steel Res.* **2018**, *147*, 203–223. [CrossRef]
23. Yang, Y.; Yang, H.; Zhang, S. Compressive Behavior of T-shaped Concrete Filled Steel Tubular Columns. *Int. J. Steel Struct.* **2010**, *10*, 419–430. [CrossRef]
24. Zuo, Z.-L.; Cai, J.; Yang, C.; Chen, Q.-J. Eccentric load behavior of L-shaped CFT stub columns with binding bars. *J. Constr. Steel Res.* **2012**, *72*, 105–118. [CrossRef]
25. Zuo, Z.-L.; Cai, J.; Yang, C.; Chen, Q.-J.; Sun, G. Axial load behavior of L-shaped CFT stub columns with binding bars. *Eng. Struct.* **2012**, *37*, 88–98. [CrossRef]
26. Zuo, Z.-L.; Cai, J.; Chen, Q.-J.; Liu, X.-P.; Yang, C.; Mo, T.-W. Performance of T-shaped CFST stub columns with binding bars under axial compression. *Thin-Walled Struct.* **2018**, *129*, 183–196. [CrossRef]
27. Zuo, Z.L.; Cai, J.; Zhu, C.H. Bearing capacity of L-shape cft stub columns with binding bars subjected to eccentric compression load. *Eng. Mech.* **2010**, *27*, 161–167+185. (In Chinese)
28. Zuo, Z.L.; Cai, J.; Liu, M.F.; Duan, W.N.; Chen, Q.-J. Experimental study of T-shaped CFT stub columns with bindingbars subjected to eccentric load. *J. Build. Struct.* **2011**, *32*, 79–89. (In Chinese)
29. Cai, J.; Zuo, Z.L.; Zhao, X.Q.; Zhu, C.H. Expermental research on eccentrically loaded L-shaped concrete-filled steel tubular stub columns with binding bars. *J. Build. Struct.* **2011**, *32*, 83–90. (In Chinese)
30. Su, G.Q. *Study on the Behavior of Cross-Shaped CFT Stub Columns with Binding Bars Subjected to Eccentric Compression*; South China University of Technology: Guangzhou, China, 2011. (In Chinese)
31. Yang, Y.; Wang, Y.; Fu, F.; Liu, J. Static behavior of T-shaped concrete-filled steel tubular columns subjected to concentric and eccentric compressive loads. *Thin-Walled Struct.* **2015**, *95*, 374–388. [CrossRef]

32. Tao, Y.; Liang, W.Q.; Zhang, S.M.; Li, X.Z.; Zhang, P. Axial compression behavior of cross-shaped concrete-filled steel tubular columns with steel bar truss stiffening. *J. Build. Struct. Suppl.* **2021**, *42*, 180–188. (In Chinese)
33. Tao, Y.; Gong, C.; Zhang, S.; Li, X.; Tan, X.; Hu, J. Axial Compressive Behavior of Cross-Shaped CFST Stub Columns with Steel Bar Truss Stiffening. *Materials* **2023**, *16*, 4147. [CrossRef] [PubMed]
34. Tao, Z.; Han, L.-H.; Wang, Z.-B. Experimental behaviour of stiffened concrete-filled thin-walled hollow steel configuration (HSS) stub columns. *J. Constr. Steel Res.* **2005**, *61*, 962–983. [CrossRef]
35. *ACI 318-11*; Building Code Requirements for Configuration Concrete and Commentary. American Concrete Institute: Farmington Hills, MI, USA, 2011.
36. *GB/T 50081-2019*; Ministry of Housing and Urban-Rural Development of the People's Republic of China, Standard for Test Method of Concrete Physical and Mechanical Properties. China Architecture & Building Press: Beijing, China, 2019. (In Chinese)
37. *GB/T 228.1-2010*; Metallic materials—Tensile Testing-Part 1: Method of Test at Room Temperature. General Adminstration of Quality Supervision, Inspection and Quarantine of China: Beijing, China, 2010. (In Chinese)
38. Zhang, S.; Guo, L.; Ye, Z.; Wang, Y. Behavior of Steel Tube and Confined High Strength Concrete for Concrete-Filled RHS Tubes. *Adv. Struct. Eng.* **2005**, *8*, 101–116. [CrossRef]
39. *GB 50936-2014*; Technical Code for Concrete-Filled Steel Tubular Structures. Architecture Industry Press of China: Beijing, China, 2010. (In Chinese)
40. *CECS 159*; Technical Specification for Structures with Concrete-Filled Rectan_Gular Steel Tube Members. China Planning Press: Beijing, China, 2004. (In Chinese)
41. *EN 1994-1-1*; Eurocode 4. Design of Composite Steel and Concrete Structures—Part 1-1: General Rules and Rules for Buildings. European Committee for Standardization (CEN): Brussels, Belgium, 2004.
42. *AISC 360-10*; Specification for Configuration Steel Buildings. American Institute of Steel Construction (AISC): Chicago, IL, USA, 2010.

*materials*

MDPI

*Article*

# Achieving Strength–Ductility Balance in TWIP Steel by Tailoring Cementite

**Zhenyu Zhao [1,2], Jian Sheng [1,2], Dazhao Li [1,2,*], Shaobin Bai [3,*], Yongan Chen [1,2], Haitao Lu [4], Pengfei Cao [1,2] and Xin Liu [1,2]**

1 School of Materials Science and Engineering, North University of China, Taiyuan 030051, China; 13903402492@163.com (Z.Z.); shengjian032@163.com (J.S.); cya894589913@163.com (Y.C.); c17824944905@163.com (P.C.); 13935140840@163.com (X.L.)
2 Shanxi Key Laboratory of Advanced Metal Materials for Special Environments, Taiyuan 030051, China
3 School of Aerospace Engineering, North University of China, Taiyuan 030051, China
4 School of Materials Science and Engineering, Taiyuan University of Science and Technology, Taiyuan 030024, China; b202314110015@stu.tyust.edu.cn
* Correspondence: dazhaoli@163.com (D.L.); sh1989bedu@163.com (S.B.)

**Abstract:** High-Mn steels are widely used in various fields. However, the FCC structure is not conducive to improving strength, limiting their development and application. In this work, hot-rolled Fe-25Mn-1Al-3Si-1C (wt.%) steel was annealed at various temperatures to tailor the cementite particles and recrystallized grains, thus achieving a balance between strength and ductility. As the annealing temperature increased from 550 to 650 °C, the volume fraction of recrystallized grains slightly increased and the volume fraction of cementite particles initially increased and then decreased, which was explained and verified by the quantitative calculation. Especially, the high-density pre-dislocation and finely dispersed cementite particles in sample AN550 resulted in a relatively low volume fraction of recrystallized grains. Interestingly, secondary deformation twinning was activated during the subsequent tensile deformation in addition to the dislocations, stacking faults, and previous deformation twinning. This complex interaction among various deformation mechanisms indued a good balance between strength and ductility, achieving an outstanding result (58.9 GPa%) regarding tensile strength and total elongation. This work offers an effective route for developing a high-Mn TWIP steel with outstanding strength–ductility balance.

**Keywords:** high-Mn TWIP steel; cementite particle; recrystallization; deformation mechanism; strength–ductility balance

## 1. Introduction

In recent years, TWIP (twinning-induced plasticity) steel has gained extensive recognition due to its favorable product of strength and elongation (PSE). It is considered a crucial candidate material in the automotive industry. The excellent PSE of steel is mainly attributed to its ductility [1–4]. The large ductility of TWIP steel is explained by its single-face-centered cubic microstructure, because numerous slip systems can be activated, and deformation twinning can be achieved during plastic deformation [5]. However, an increase in ductility will inevitably lead to a decrease in strength. Therefore, it is important to optimize the strength–ductility balance of TWIP steel, enhancing its strength.

Numerous scholars have employed various approaches to enhance the strength of TWIP steel. The yield strength of steel can be enhanced through the use of dislocation, fine grains, precipitates, etc. [6]. Early work on the micro-alloying of TWIP steels was employed to enhance strength via solid solution strengthening and the Orowan precipitate bypassing mechanism [7]. Liang et al. [8] enhanced the yield strength (YS) of high-Mn steel through Nb micro-alloying. However, this method not only resulted in high costs but also increased the difficulty of smelting [9]. Some researchers improved mechanical properties by introducing nano-scale precipitates. Qiu et al. [5] employed a cold-roll annealing process to refine the grains and introduce a uniform distribution of nano-scale precipitates (κ-carbide), which interact with the dislocation and effectively improve the strength of high-Mn steel. However, the cold-rolling process leads to a dislocation density approaching saturation, limiting continuous work hardening during deformation, thus leading to a significant deterioration in ductility [5]. Furthermore, the cold-rolling process has drawbacks, such as high energy consumption and complex procedures. The hot-rolling process requires lower rolling forces and offers a lower cost. In addition, the surface quality and strength of hot-rolled steel plates have been improved significantly in recent years [9]. Studies have shown that the combination of nano-scale precipitates and dislocations may reduce the mutual limitations in terms of strength and ductility [5]. At the same time, the introduction of partially recrystallized microstructures is conducive to improvements in ductility [10]. Hu et al. [11] pointed out that the precipitation of nano-scale precipitates has a significant impact on recrystallization behavior via solute drag and Zener pinning effects. In turn, recrystallization interacts with precipitates, thereby influencing mechanical properties. Ren et al. [12] reported that tailoring the precipitation of nano-scale carbides via the hot-rolling–annealing process and mixing the recrystallization structure can improve strength while retaining ductility. The results show that this is an effective method with which to increase strength by introducing carbides and recrystallization structures via the hot rolling-annealing process. However, the influence of nano-scale carbides on recrystallization behavior remains unclear during annealing. In addition, the existence of precipitates, dislocations, partial recrystallization, and other microstructures results in complex deformation mechanisms. Meanwhile, the reaction mechanism of competition and coupling during plastic deformation remains unclear.

In this work, the hot-rolling process is employed to refine grains and introduce high dislocation densities in Fe-25Mn-1Al-3Si-0.2C (wt.%) steel. Then, carbide precipitation is tailored by controlling the annealing temperature and achieving an excellent balance between strength and toughness. The PSE value reaches 58.9 GPa%. Subsequently, the influence of precipitates on recrystallization behavior is discussed. Moreover, the evolution behavior of microstructure and various deformation mechanisms during tensile deformation are investigated.

## 2. Experimental Section

The composition of the TWIP steel used in the experimental methods is Fe-25Mn-1Al-3Si-0.2C (wt.%). This was melted in a 50 kg vacuum induction furnace and then cast into a mold with dimensions of Ø170 × 350 mm. After homogenizing at 1200 °C for 2 h in an Ar atmosphere, the ingot was hot-rolled into 3 mm plates in 7 passes along the axial direction, with initial and final forging temperatures of 1150 °C and 900 °C, respectively. Large deformations were applied during the hot-rolling process to induce high dislocation densities and fine grains. After hot rolling, the specimens were cut from the rolled plate and then subjected to a modulated annealing treatment in an Ar atmosphere at 550 °C, 600 °C, and 650 °C for 30 min, respectively. The test samples were designated as HR (hot-rolled state), AN550, AN600, and AN650. The process flow diagram is shown in Figure 1.

The microstructures of the specimens were characterized by X-ray diffraction (XRD, SmartLab-3 KW), scanning electron microscopy (SEM), electron backscatter diffraction (EBSD), and transmission electron microscopy (TEM). To determine the existing phases,

X-ray diffraction was performed by scanning at 40 kV with a current of 200 mA using CuKα radiation in the 2θ range of 20° to 100° at a scan rate of 5°/min. The microstructure was observed in the transverse direction (perpendicular to the rolling surface) using a field-emission SEM (Tescan Mira3, TESCAN Group, Shanghai, China) equipped with an EBSD system. The SEM-EDS result was analyzed by using EDAX TEAM. The SEM and EBSD samples were first mechanically polished and thereafter electro-polished for 35–40 s using 10% perchloric acid and 90% ethanol electrolyte at 15.8 V and −20 °C. The EBSD scanning step size was 0.4 μm, and the size of the scanned are was 291.6 μm × 223.69 μm. The data were analyzed utilizing OIM7.3 software. Cleaning procedures were implemented. We appropriately employed the Grain CI standardization cleanup method. For this method, the parameter minimum grain size was set to 1, and the parameter grain tolerance angle was set to 5 in OIM7.3 software. The TEM sample was initially mechanically ground to a thickness of 50 μm and then punched into 3 mm diameter disk. Subsequently, a twin-jet electric polisher was employed for further thinning in a solution containing 5% perchloric acid and ethanol. Finally, TEM observation was conducted using the JEM-2100F (JEOL, Tokyo, Japan) field-emission electron microscope at an acceleration voltage of 200 kV. The tensile samples, 70 mm long and 6.0 mm wide, were made by electrical discharge machining (EDM) of the hot-rolled plate along the rolling direction. Tensile testing was performed on a UTM 5305 tester with a strain rate of $10^{-3}$ $s^{-1}$. Tensile experimental data were obtained three times for each sample, and the average value was determined.

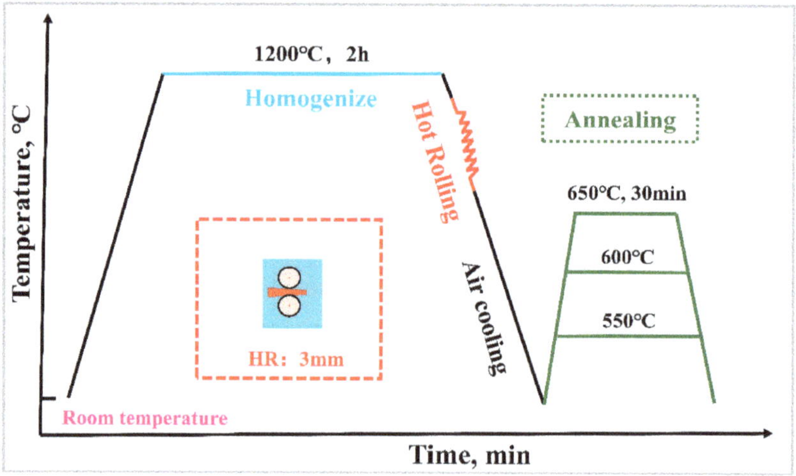


**Figure 1.** Abstract process flow diagram.

## 3. Results

### 3.1. Microstructure Characterization

Figure 2 shows the microstructures of samples under various conditions. In the hot-rolled sample, a large number of annealing twins and slip bands were observed in austenite grains (Figure 2a). However, no significant precipitates were observed (Figure 2a). Figure 2b–d show the microstructures of samples AN550, AN600, and AN650, respectively. For all annealed samples, the annealing twins and precipitates are present within austenite grains and along grain boundaries, respectively. In contrast, the precipitate size in sample AN550 was much smaller than in other samples. As shown in the enlarged image in Figure 2b, numerous fine precipitates were intra-granular in addition to the grain boundaries (GBs). For samples AN600 and AN650, the increase in annealing temperature inevitably induced the coarsening of precipitates (green ellipses), especially the particles in grain boundaries. Figure 2e shows the EDS result of the precipitate in sample AN600. Its chemical composition was 38.7Fe-18Mn-16.3C (at%), implying that similar precipitates included (Fe, Mn)$_3$C cementite particles. Figure 2f displays the XRD results of samples before and after annealing. It can be seen that only single γ-Fe diffraction peaks were exhibited, indicating a fully austenitic single phase without any phase transformation.

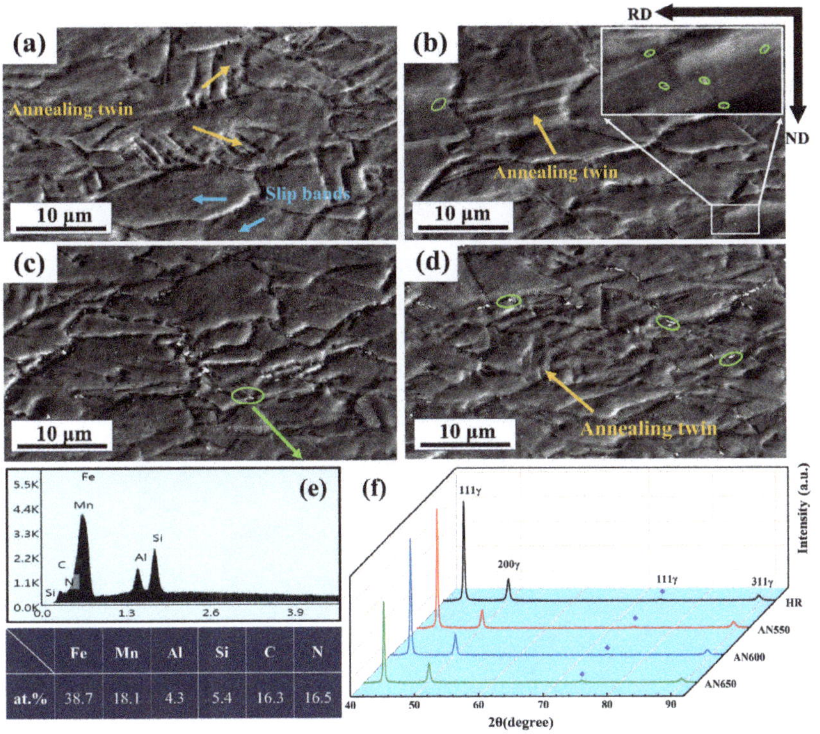


| | Fe | Mn | Al | Si | C | N |
|---|---|---|---|---|---|---|
| at.% | 38.7 | 18.1 | 4.3 | 5.4 | 16.3 | 16.5 |

**Figure 2.** SEM microstructures of samples: (**a**) HR; (**b**) AN550; (**c**) AN600; (**d**) AN650. (**e**) EDS result of cementite particles in sample AN600. (**f**) XRD spectra of samples under various conditions.

EBSD and TEM were performed to analyze the microstructures (Figure 3). Banded structures (dashed lines) and precipitates (M$_3$C) were present within the austenitic matrix. These banded structures were formed due to the uneven distribution of Mn, which also affected recrystallization behavior [13]. For the hot-rolled sample, a small number of fine precipitates were detected (Figure 3a). In contrast, there were many more nano-scale particles exhibited in AN550 samples (Figure 3b). TEM-EDS mapping (Figure 3f) clearly showed that Fe, Mn, and C were enriched in the precipitates. The corresponding selected area diffraction pattern (SADP) verified these nano-scale particles were (Fe, Mn)$_3$C cementite (Figure 3e). As the annealing temperature increased, the fine precipitates within grains gradually dissolved, while the precipitates distributed along GBs gradually coarsened (Figure 3c,d).

This finding was highly consistent with the SEM results in Figure 2. More than five different SEM images of the sample were used to measure the volume fraction and average size of cementite particles using Image-Pro software6.0. The volume fraction ($f$) and average size ($x$) of cementite particles were statistically measured (Figure 4).

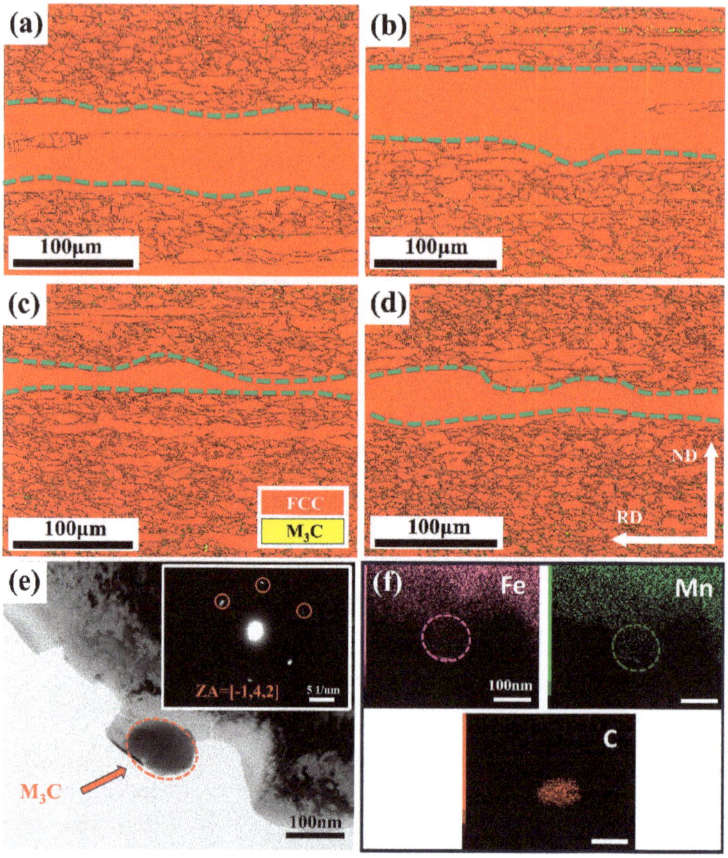


**Figure 3.** EBSD and TEM micrographs of samples under various conditions. (**a**–**d**) The phase diagrams of coexisting austenite and precipitate: HR, AN550, AN600 and AN650, respectively. (**e**) shows the precipitates observed in the AN550 sample, and (**f**) is the EDS map of the red region of (**e**).

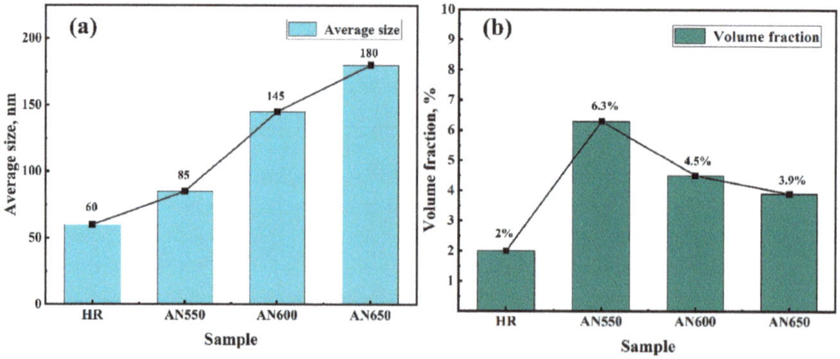


**Figure 4.** (**a**,**b**) are the volume fraction and average grain size of the carbides, respectively.

Figure 5a–d show the kernel average misorientation (KAM) maps and the corresponding KAM values of all samples. The KAM value can reflect the geometric dislocation density and recrystallization volume fraction of the sample [13–16]. The KAM value decreased from 1.22 to 0.98 with the increased annealing temperature. It can be seen that the recrystallized volume fraction and the dislocation density increase and decrease with the increase in annealing temperature, respectively.

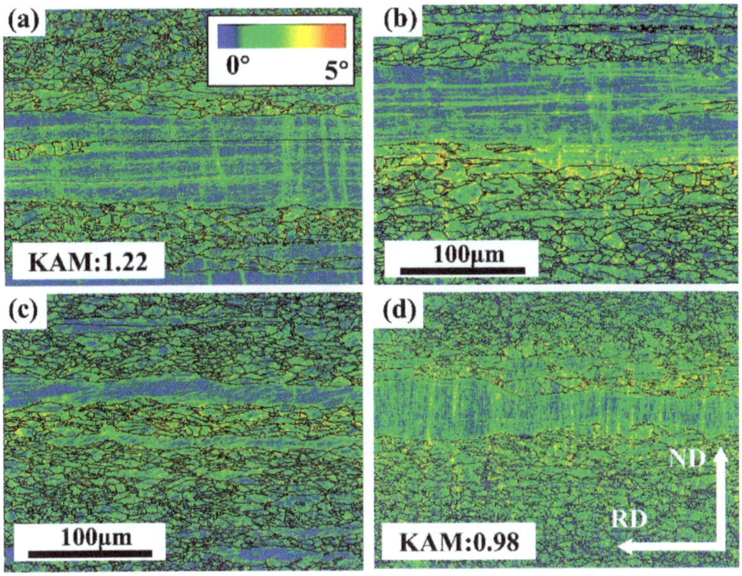


**Figure 5.** KAM maps of the samples under various conditions. (**a**) HR, (**b**) AN550, (**c**) AN600, (**d**) AN650. 0–1°, recrystallized grains; 1–5°, non-recrystallized grains.

The specific KAM values calculated are shown in Figure 6a. Figure 6b shows the statistical distribution of grain sizes for all samples, which were fitted with a normal distribution function and are summarized in Table 1. It can be seen that the grain size decreases with the increase in annealing temperature.

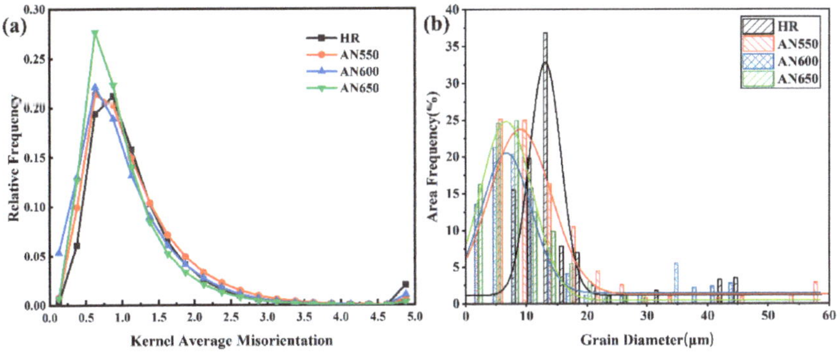


**Figure 6.** (**a**) KAM distribution map of the sample. (**b**) Average grain size distribution of the sample under various conditions (excluding Mn segregation band).

**Table 1.** The volume fraction and average grain ($d$)/particle ($x$) size in the samples.

| Date | HR | AN550 | AN600 | AN650 |
|---|---|---|---|---|
| Average particle size($x$)/nm | 60 | 85 | 145 | 180 |
| Volume fraction($f$) | 0.020 | 0.063 | 0.045 | 0.039 |
| Average grain size(d)/μm | 15.8 | 13.8 | 10.8 | 8.5 |

### 3.2. Mechanical Properties

Figure 7a,b show the engineering stress–strain curve and the measured mechanical properties of all samples. As the annealing temperature increased, the yield strength (YS) and ultimate tensile strength (UTS) gradually decreased. Then, the total elongation (TE) and UTS × TE (PSE) showed a trend of first increasing and then decreasing. The largest elongation of 54.9% and the highest PSE of 58.9 GPa% were obtained for sample AN550.

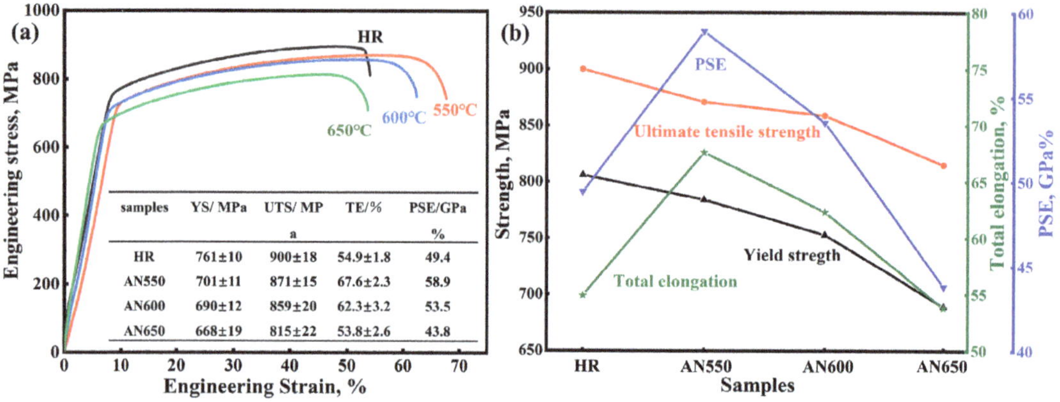


**Figure 7.** Mechanical responses at different annealing temperatures. (**a**) Engineering stress–strain curves; (**b**) the tensile mechanical properties of samples in terms of yield strength, ultimate tensile strength, total elongation, and UTS × TE.

## 4. Discussion

### 4.1. Strengthening Mechanism

The yield strength of TWIP steels is influenced by lattice friction stress ($\sigma_f$) and multiple strengthening mechanisms, such as the solid solution strengthening ($\sigma_s$), grain refinement strengthening ($\sigma_g$), dislocation strengthening ($\sigma_d$), and precipitation strengthening ($\sigma_p$). Due to the higher content of replacement atoms in high-manganese TWIP steels, the contribution of $\sigma_s$ to the yield strength increment is less than 50 MPa [17–19]. Therefore, the $\sigma_s$ value is neglected here. The contribution of various strengthening mechanisms to the yield strength can be calculated using Equation (1):

$$\Delta\sigma_y = \sigma_f + \sigma_p + \sigma_g + \sigma_d \tag{1}$$

where $\sigma_f$ is 250.7 MPa [5,20]. The contribution of $\sigma_p$ to yield strength can be calculated using the following equation by simple calculation [21]:

$$\sigma_p = \frac{0.538\,Gbf^{\frac{1}{2}}}{x}\ln\left(\frac{x}{2b}\right) \tag{2}$$

where G is the shear modulus (72.0 GPa); b is the Burgers vector (0.254 nm); and $f$ and $x$ represent the volume fraction and average size of precipitate, respectively (Table 1). The measured $\sigma_p$ values of samples AN550, AN600, and AN650 are 147, 83, and 66 MPa, respectively.

The contribution of $\sigma_g$ to yield strength can be estimated using the Hall–Petch equation [19]:

$$\sigma_g = k_y d_i^{-\frac{1}{2}} \tag{3}$$

where $k_y$ is constant, with a value of ~683.07 MPa $\mu m^{1/2}$ [5], and $d_i$ represents the average grain size. The calculated $\sigma_g$ values for samples AN550, AN600, and AN650 are 185, 207, and 232 MPa, respectively.

According to Ref. [21], The contribution of $\sigma_d$ can be used in the Taylor hardening law equation for evaluation.

$$\sigma_d = M\alpha Gb\sqrt{\rho} \tag{4}$$

where M is the Taylor factor of 3.06 for FCC materials, $\alpha$ is the property constant of 0.25, and $\rho$ is the dislocation density [18]. The average dislocation density ($\rho$) was calculated using Equation (5) [21]:

$$\rho = \frac{\beta^2}{4.35 \times b^2} \tag{5}$$

where $\beta$ is the full width at half maximum (FWHM) of (111)$\gamma$ diffraction peaks from X-ray diffraction patterns. The measured $\rho$ value gradually decreased from $0.86 \times 10^{14}$ in sample AN550 to $0.45 \times 10^{14}$ in sample AN650 (Table 2). Subsequently, the calculated $\sigma_d$ values for samples AN550, AN600, and AN650 were 129, 115, and 93 MPa, respectively.

**Table 2.** Full width at half maximum examined along the RD and the dislocation density for austenite.

| Samples | FWHM (rad) | P ($m^{-1} \times 10^{14}$) |
|---|---|---|
| | γ(111) | |
| AN550 | 0.004905946 | 0.86 |
| AN600 | 0.004355295 | 0.68 |
| AN650 | 0.003540051 | 0.45 |

Figure 8 shows the contribution of various strengthening mechanisms to the yield strength increments. The results show that the calculated values are similar to the experimental values. As the annealing temperature increased, the occurrence of recovery and the coarsening and dissolving of precipitate induced the decreases in ρ and $\sigma_d$, respectively, while the activation of recrystallization further refined the grain and led to the increase in $\sigma_g$.

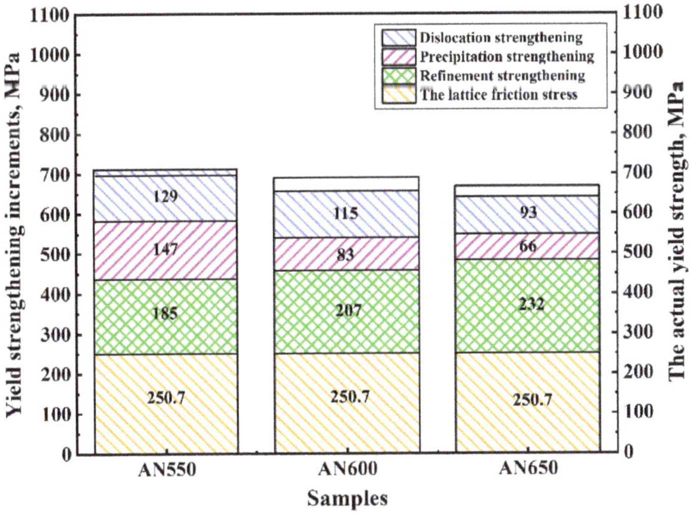


**Figure 8.** The contribution of the individual strengthening mechanism and the experimental yield strength.

*4.2. Precipitates Affect Recrystallization Behavior*

Figure 9 shows the results of recrystallized microstructure analysis. Meanwhile, the red, yellow, and blue colors represent the deformed, partially recrystallized, and fully recrystallized grains, respectively. The fraction and size of recrystallized grain gradually increased with the increasing annealing temperature, while the average substructured grain size significantly decreased. This phenomenon was especially significant in sample AN650.

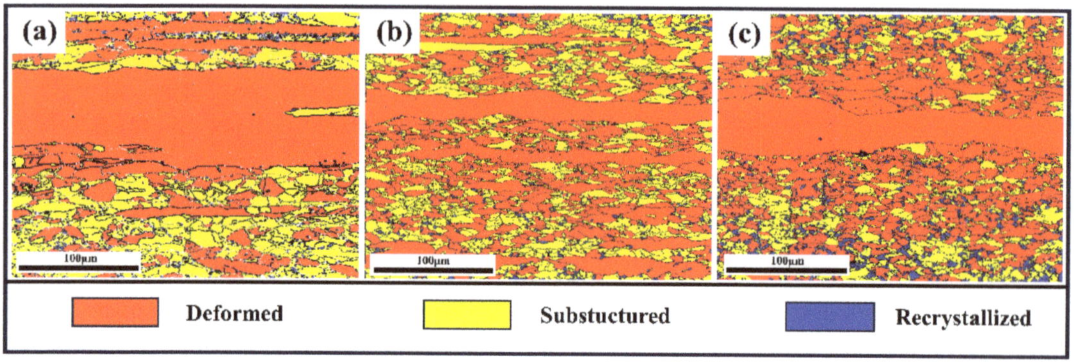


**Figure 9.** Recrystallized distribution maps of samples AN550 (**a**), AN600 (**b**), and AN650 (**c**), respectively. The red, yellow, and blue colors represented the deformed, partially recrystallized, and fully recrystallized grains, respectively.

Simultaneous deformation during hot rolling leads to dense dislocation residues [5]. After annealing at a relatively lower temperature (550 °C), a relatively higher density of pre-existing dislocations (pre-dislocation) was retained in sample AN550. Due to the loose arrangement of atoms in the pre-dislocation region, the diffusion rate of atoms is much higher than that in the regular lattice, resulting in a faster atom diffusion rate [22]. The diffusion of atoms in the lattice is a thermally activated process that can be strongly accelerated using defects such as dislocations. This is ascribed to a reduced activation barrier [23]. The formation of cementite particles requires the diffusion of component elements at these dislocations through a pipeline mechanism [24,25]. Therefore, a higher pre-dislocation density in sample AN550 can ensure a faster solute diffusion rate and further provide more nucleation points for the cementite particles. In addition, the pre-dislocation structure includes numerous tangle dislocations, which increases the stability of the pre-dislocation structure. Subsequently, partial pre-dislocations recover during the annealing process. However, the early precipitated cementite particles can pin the pre-dislocations and hinder the recovery of dislocation, which further improves the stability of the dislocation structure. Specifically, fine carbides lead to the elimination of regions of high lattice misorientation, which are preferred sites for recrystallization nucleation, thus inhibiting recrystallization [24]. In addition, the presence of solute atoms and fine cementite particles hinders dislocation motion within the grain, thereby pinning sub-grain boundaries and preventing nucleation. Simultaneously, intergranular cementite particles impede grain boundary migration, thereby suppressing recrystallization, entailing solute drag and Zener pinning effects caused by precipitation [11,26]. This is because the diffusion rate of solute elements is lower than that of matrix atoms, resulting in solute atoms at the grain boundaries obstructing grain boundary migration and inhibiting grain growth [11]. Meanwhile, the finely dispersed precipitates provide greater resistance to grain boundary migration, leading to a more pronounced pinning effect and consequently reducing the rate of GB migration. The pinning pressure can be expressed by Equation (6) [27]:

$$Pz = \frac{3F_v\gamma}{2r} \tag{6}$$

where $Pz$ is the pinning pressure, $Fv$ and r are the volume fraction and size of the precipitate, respectively, and $\gamma$ is the boundary surface energy. The relationship between $P_z$ and $Fv/r$ are observed to be proportional. Sample AN550 contains more fine precipitates (Figure 2b). According to Table 1, the $Fv/r$ values of samples AN550, AN600, and AN650 are 0.074, 0.031, and 0.022, respectively. Therefore, the finely dispersed precipitates exert greater resistance to GB migration in the AN550 sample. In addition, the shape of the precipitates also affects the pinning effect [11]. It was reported that the spherical precipitates were able to provide a higher pinning pressure [24].

In contrast, samples AN600 and AN650 were annealed at relatively high temperatures (600 °C and 650 °C). With the increase in annealing temperature, the continuous occurrence of recrystallization led to a decrease in dislocation density and a decrease in nucleation sites. However, a higher annealing temperature accelerates the diffusion of solute atoms to defects, and the solute atoms gather and even cause segregation at the GBs, leading to the coarsening and even dissolution of precipitates [26]. The consumption of solute atoms and the coarsening of precipitates weakens the drag effect and pinning effect [26]. The pinning pressure at GBs is specified in Equation (7) [28]:

$$P_{GB} = \frac{F_{V_{GB}}\gamma D}{4r^2} \tag{7}$$

where $F_{VGB}$ and D are the volume fraction and size of carbides at grain boundaries, respectively. $P_{GB}$ is proportional to the $F_{VGB}/D$ ratio. According to Table 1, it can be inferred that an increase in annealing temperature leads to a gradual reduction in the pinning effect observed in AN600 and AN650 samples, thereby weakening the resistance to grain boundary migration and facilitating recrystallization (Figure 9b,c). In addition, it was reported that recrystallization preferentially nucleates at high orientation differences [24]. With the increase in annealing temperature, precipitates gradually dissolve and segregate, resulting in the uneven precipitation of carbides. The coarse particles interact with dislocation to provide the orientation gradient required for nucleation, which further facilitates recrystallization.

### 4.3. Deformation Mechanism and Strain-Hardening Behavior

The strain-hardening behavior was closely related to the deformation mechanism. Figure 10 shows the work-hardening rate-true strain curves and the corresponding fracture morphologies of all samples. All curves can be divided into three stages (I, II, and III). At stage I, the work hardening rate decreased rapidly due to the dynamic recovery of dislocations [14]. Then, the work hardening rates gradually reached constant values with the increasing true strain (stage II). Finally, the work-hardening rate significantly decreased until fracturing (stage III). The fracture morphologies of samples are shown in the top right corner of Figure 10a–d. All samples showed a large number of dimples. In contrast, the dimples were fine and evenly distributed in sample AN550 (Figure 10b), leading to the typical features of ductile fractures [14,29].

Figure 11 shows the interrupted tensile microstructures of sample AN550 under various true strains. Recrystallization was inhibited by the precipitation of cementite in sample AN550 (Figure 9a). Before deformation, a high dislocation density was observed along both sides of the annealing twins, while the dislocation density inside was lower (Figure 11a).

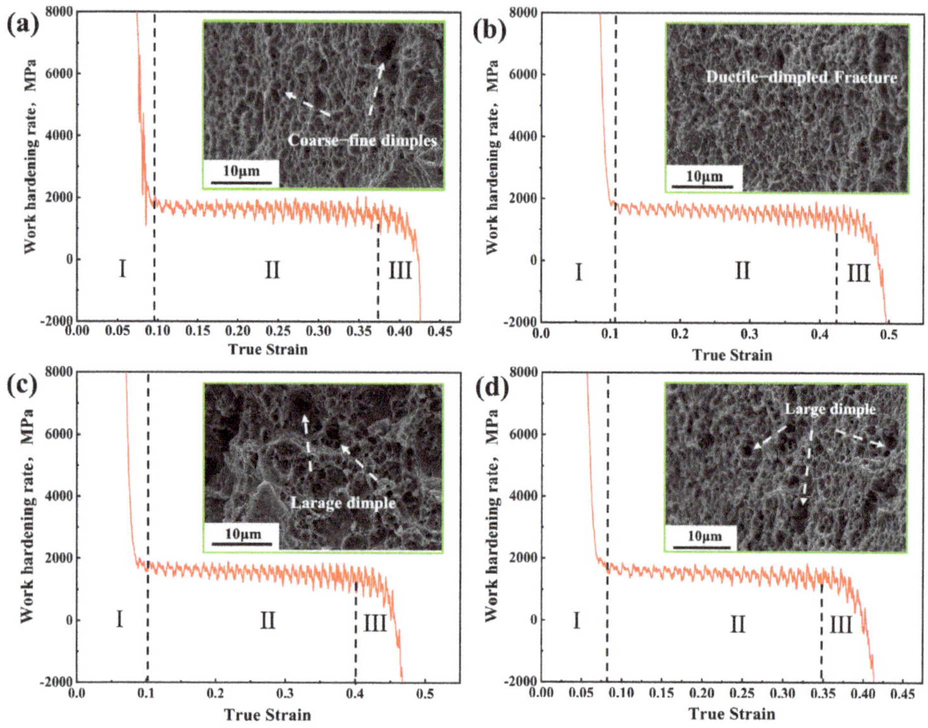


**Figure 10.** Work hardening rate (WHR) and fracture morphologies of annealed samples: (**a**) sample HR; (**b**) sample AN550; (**c**) sample AN600; (**d**) sample AN650.

At a true strain of 0.15, a number of dislocations were activated. The interaction between dislocations gradually forms dislocation tangles and several dislocation substructures, such as dislocation cells, as shown in Figure 11b. Simultaneously, a large number of dislocations accumulate at the GBs, due to the blocking effect of GBs, resulting in stress concentrations. In order to achieve release stress, grain boundaries become a source of dislocation radiation, resulting in rapid dislocation proliferation [30]. It was reported that the dissociation of pre-dislocation from partial dislocation can provide nucleation points for deformation twins and promote the formation of stratification faults [31]. Shockley partial dislocations were emitted to further promote the formation of stacking faults and twins at the GBs, enhancing WHR [30], as shown in Figure 11c.

As the strain increases to 0.3, more dislocations are activated. A number of dislocations accumulate at the GBs, so that the stress reaches the critical stress required for the formation of twinning, thus promoting the generation of twins. In addition, the presence of a large number of dislocations moving near the stacking faults contributes to the formation of deformation twins [31]. Therefore, with the increase in strain, a large number of primary deformation twins are formed, causing continuous work hardening. Simultaneously, the interaction between stacking fault and deformation twin promotes the formation of the Lomer–Cottrell lock (Figure 11d), thus hindering the dislocation movement and enhancing the WHR [12,31]. In addition, the twin boundaries hinder dislocation motion, causing dislocation tangles. Further movement of the dislocation results in the formation of dislocation substructures between the twin lamellae (Figure 11e,f). However, the dislocation substructures can further impede dislocation movement, which promotes work hardening [23]. As strain increases, the formation of secondary twins can hinder dislocations from crossing the grain boundary barrier and avoid their annihilation, thus increasing dislocation density. Meanwhile, numerous deformation twins play a crucial role in grain refinement through the dynamic Hall–Petch effect (Figure 11g–i), thereby reducing the average free path of dislocations and further enhancing the WHR [1,32].

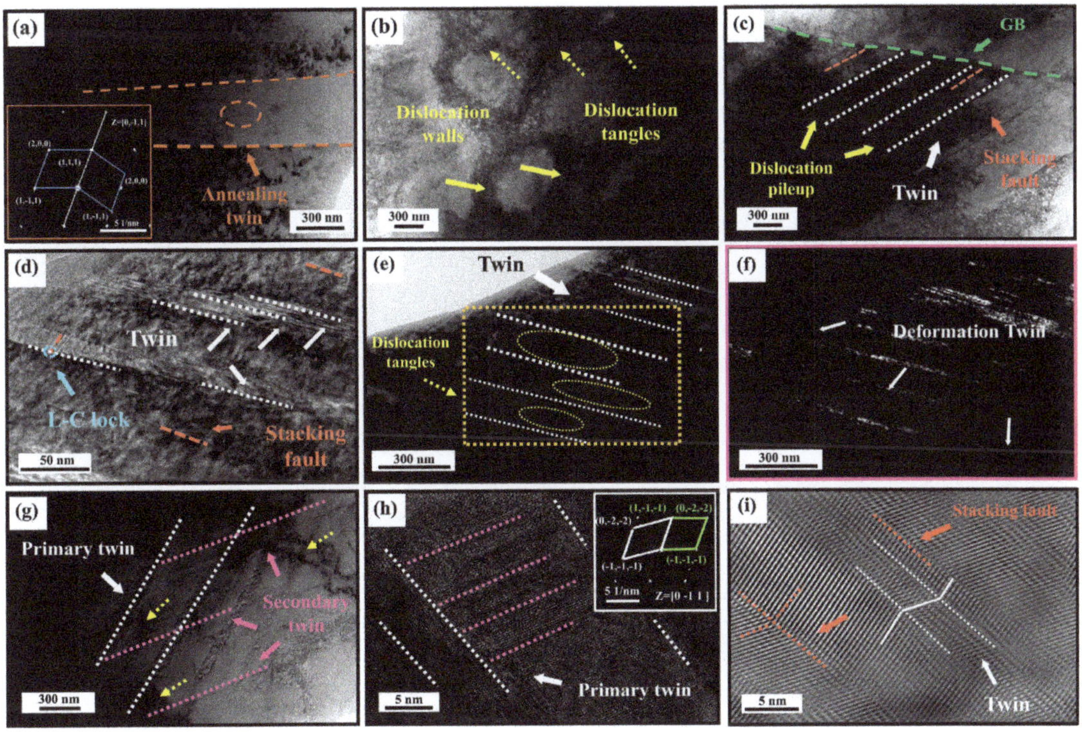


**Figure 11.** The interrupted tensile microstructures of sample AN550 under various true strains. (**a**) Before deformation. (**b**,**c**) At a true strain of 0.15, (**b**) dislocation tangles and dislocation walls; (**c**) stacking faults and twins near GB. (**d**–**f**) At a true strain of 0.3, (**d**) twin and L-C lock; (**e**,**f**) are the bright and dark field images, respectively. (**g**–**i**) After fracture, (**g**,**h**) are the bright field and HRTEM images of the initial and secondary twins, respectively; (**i**) is an inverse fast Fourier transform (IFFT) image.

Figure 12a–c show the TEM images of sample AN650 before and after tensile deformation. The increase in annealing temperature weakens the pinning effect, thus promoting recrystallization and leading to a low dislocation density in the recrystallization region (Figures 9c and 12a). The large number of recrystallized grains provides abundant space for the multiplication and accumulation of dislocations during subsequent plastic deformation [26]. As the strain increases, most of the strain is distributed to the softer recrystallized regions. To accommodate the strain gradient, long-range geometrically necessary dislocations (GNDs) are generated, thereby avoiding strain localization. Meanwhile, continuous back stress hardening is provided, which contributes to the WHR [23]. In addition, precipitate particles strongly inhibit dislocation movement during the straining process, further increasing WHR (Figure 12b).

However, the ductility of samples AN600 and AN650 deteriorated compared with sample AN550. An increase in annealing temperature induces the coarsening of precipitates, especially the particles in GBs (Figure 2a–c). On one hand, the large-sized carbides at the interface can increase the stacking fault energy at GBs, which increases the critical stress for twin generation, thereby lowering ductility. On the other hand, the coarse intergranular precipitate changes the lattice parameters of austenite. It is easy to produce stress concentrations to form vacancies that evolve into crack sources [33–35], further lowering ductility. In addition, the uneven precipitation of coarse carbides can cause local stress concentrations due to the pile-up of slip bands, further compromising ductility.

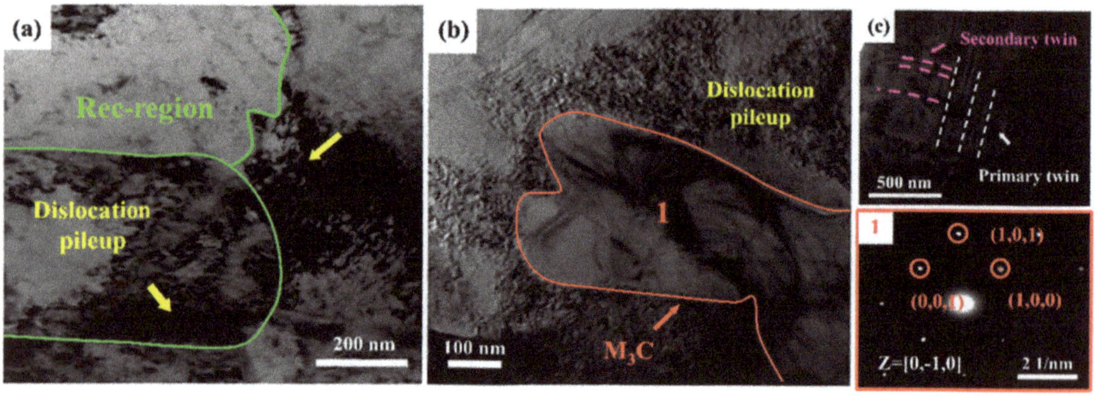


**Figure 12.** TEM images of sample AN650. (**a**) represents the microstructure prior to deformation, which displays a recrystallized structure; (**b**,**c**) are the microstructures after fracture; (**b**) shows the interaction between $M_3C$ and dislocations; (**c**) shows primary twins and secondary twins.

*4.4. Microstructural Evolution*

Figure 13 displays the abstract microstructure evolution of annealed samples before and after tensile deformation. With the increase in annealing temperature, the size of cementite particles gradually coarsened in samples HR, AN550, AN600, and AN650. The volume fraction of cementite particles first increased and then decreased. The volume fraction of recrystallization gradually increased. In contrast, sample AN550 has the highest volume fraction of cementite and a relatively high dislocation density (Figure 13a–d).

In the process of tensile deformation, multiple deformation mechanisms were generated (Figure 13e,f). The high-density dislocations activate the TWIP effect in all samples. As the dislocations in the recrystallization zone are relatively even, they facilitate the formation of thinner deformation twins. With the increase in tensile strain, secondary twins are activated, and the formation of substructures is promoted (Figure 13e,f). As the strain must be continuous at the connection interface, GNDs are produced to ease the strain gradient and avoid the formation of strain gradients in the vicinity of the interface [23]. In the AN550 sample, the volume fraction of partial recrystallization is elevated. Meanwhile, the pinning effect of an appropriate quantity of cementite particles with relatively small sizes on grain boundaries dislocations is significantly increased, which significantly promotes the dislocation storage energy (Figure 13e). For sample AN650, even though the volume fraction of recrystallization is further enhanced, a large number of coarsened cementite particles are distributed along the grain boundaries, resulting in the deterioration of ductility.

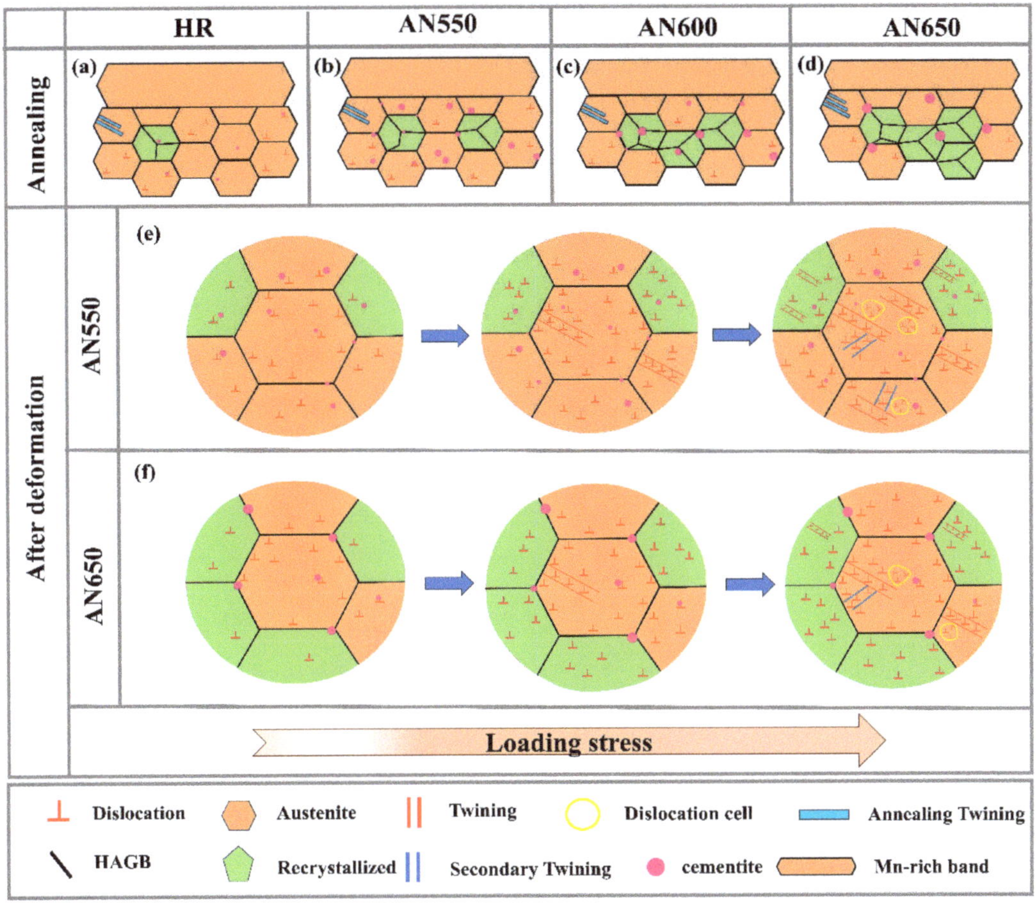


**Figure 13.** Microstructural evolution diagrams of different samples after annealing (**a**–**d**), and after deformation for AN550 and AN650 samples (**e**,**f**). (**a**) HR; (**b**,**e**) AN550; (**c**) AN600; (**d**,**f**) AN650.

## 5. Conclusions

In this study, the hot-rolling annealing process tailors the precipitation of cementite particles. Cementite influences recrystallization behavior. The conclusions are as follows:

1. Higher-density pre-dislocation promotes the precipitation of fine cementite particles, resulting in higher pinning pressure. With the increase in annealing temperature, the coarsening of cementite particles weakens the pinning pressure.
2. An appropriate amount of fine cementite particles results in a relatively low volume fraction of recrystallization in sample AN550, which shows comprehensive mechanical properties with a YS of 701 MPa and a PSE value of 58.9 GPa%. Precipitation and dislocation strengthening play an essential role in YS.
3. Multiple deformation mechanisms of dislocation, stratification, primary twin, and secondary twin interact with each other, providing continuous work hardening for sample AN550.

**Author Contributions:** Z.Z.: investigation, methodology, writing—original draft. J.S.: data accumulation. D.L.: conceptualization, supervision, writing—review and editing. S.B.: investigation, writing—review and editing. Y.C.: methodology. H.L.: investigation. P.C.: data accumulation, methodology. X.L.: performing experiments. All authors have read and agreed to the published version of the manuscript.

**Funding:** This work was financially supported by National Key Laboratory Foundation of Science and Technology on Materials under Shock and Impact (No. 6142902240502), the National Natural Science Foundation of China (Nos. 52375392, 52205402, 52071300) and the Fundamental Research Program of Shanxi Province (CN) (No. 202203021212318).

**Institutional Review Board Statement:** Not applicable.

**Informed Consent Statement:** Not applicable.

**Data Availability Statement:** The original contributions presented in this study are included in the article. Further inquiries can be directed to the corresponding authors.

**Conflicts of Interest:** The authors declare that we have no known competing financial interests and personal relationships that could have appeared to influence the work reported in our work.

## References


1. Bai, S.B.; Chen, Y.A.; Liu, X.; Lu, H.H.; Bai, P.K.; Li, D.Z.; Huang, Z.Q.; Li, J.Y. Research Status and Development Prospect of Fe-Mn-C-Al System Low-Density Steels. *J. Mater. Res. Technol.* **2023**, *25*, 1537–1559. [CrossRef]
2. Zhi, H.; Antonov, S.; Zhang, C.; Guo, Z.; Su, Y. Origins of back stress strengthening in Fe-22Mn-0.6C (-3Al) TWIP steels. *Mater Sci. Eng. A* **2020**, *792*, 139834. [CrossRef]
3. Olugbade, T.O.; Oladapo, B.I.; Ting, T.T. Superior strength and wear resistance of mechanically deformed High-Mn TWIP steel. *Colloids Surf.* **2024**, *696*, 134388. [CrossRef]
4. Jia, A. Progress and prospects of natural gas development technologies in China. *Nat. Gas Ind. B* **2018**, *5*, 547–557. [CrossRef]
5. Qiu, X.; Wang, Y.; Yang, B.; Xiong, Z.; Cheng, X. Multiphase precipitation behavior and tensile properties of a Fe-Mn-Al-Mo-C austenitic lightweight steel. *Mater Sci. Eng. A* **2023**, *885*, 145654. [CrossRef]
6. Su, S.S.; Seokmin, H.; Junghoon, L.; Byeong-Chan, S.; Sung-Kyu, K.; Byeong-Joo, L.; Nack, J.K.; Sunghak, L. Effects of Mn and Al contents on cryogenic-temperature tensile and Charpy impact properties in four austenitic high-Mn steels. *Acta Mater.* **2015**, *100*, 39–52.
7. Jun, G.H.; Kyung, K.J.; Bian, J.; Hardy, M.; Taejin, S.; Kyu, K.S.; Cooman, D. Partially-recrystallized, Nb-alloyed TWIP steels with a superior strength- ductility balance. *Mater Sci. Eng. A.* **2018**, *711*, 130–139.
8. Liang, Z.; Huang, M.; Yen, H.; Scott, C. On the Work-hardening Mechanism of TWIP Steels Strengthened by Nanometre-sized Vanadium Carbides. In Proceedings of the 2nd International Symposium on Automobile Steel, Anshan, China, 21–24 May 2013; China Machine Press: Beijing, China, 2014; pp. 113–117.
9. Shu, Y.; Tianle, L.; Taosha, L.; Qi, C.J.; Yang, Z. By controlling recrystallization degree: A plain medium Mn steel overcoming Lüders deformation and low yield-to-tensile ratio simultaneously. *Mater Sci. Eng. A* **2019**, *758*, 79–85.
10. Lee, H.; Jo, M.C.; Sohn, S.S.; Zargaran, A.; Ryu, J.H.; Kim, N.J.; Lee, S. Novel medium-Mn (austenite + martensite) duplex hot-rolled steel achieving 1.6 GPa strength with 20% ductility by Mn-segregation-induced TRIP mechanism. *Acta Mater.* **2018**, *147*, 247–260. [CrossRef]
11. Hu, G.W.; Zeng, L.C.; Du, H.; Wang, Q.; Fan, Z.T.; Liu, X.W. Combined effects of solute drag and Zener pinning on grain growth of a NiCoCr medium-entropy alloy. *Intermetallics* **2021**, *136*, 107271. [CrossRef]
12. Ren, J.K.; Chen, Q.Y.; Chen, J.; Liu, Z.Y. Enhancing strength and cryogenic toughness of high manganese TWIP steel plate by double strengthened structure design. *Mater Sci. Eng. A* **2020**, *786*, 139397. [CrossRef]
13. Zhang, J.Y.; Xu, Y.B.; Wang, Y.; Wang, S.Y. Achieving heterogeneous TWIP/TRIP steel with 1GPa strength and 72% ductility by introducing austenite+martensite lamella structure. *Mater. Charact.* **2023**, *197*, 112709. [CrossRef]
14. Yin, F.X.; Xia, H.; Feng, J.H.; Cai, M.H.; Zhang, X.; Sawaguchi, T. Mechanical properties of an Fe-30Mn-4Si-2Al alloy after rolling at different temperatures ranging from 298 to 1073 K. *Mater. Sci. Eng. A* **2018**, *725*, 127–137. [CrossRef]
15. Liu, J.X.; Li, L.L.; Yang, S.W.; Ding, C.; Wang, E.; Yu, X.P.; Wu, H.B.; Niu, G. Effect of intragranular κ carbides and intergranular precipitates on the hot deformation mechanism and dynamic recrystallization mechanism of Fe–28Mn–11Al–1.5C–5Cr lightweight steel. *J. Mater. Res. Technol.* **2023**, *27*, 2346–2362. [CrossRef]
16. Sohn, S.S.; Song, H.; Min, C.J.; Song, T.; Lee, S. Novel 1.5 GPa-strength with 50%-ductility by transformation-induced plasticity of nonrecrystallized austenite in duplex steels. *Science* **2017**, *7*, 1255.

17. Niu, G.; Zurob, H.S.; Misra, R.D.K.; Wu, H.B.; Zou, Y. Strength-ductility synergy in a 1.4 GPa austenitic steel with a heterogeneous lamellar microstructure. *J. Mater. Sci. Technol.* **2022**, *106*, 131–138. [CrossRef]
18. Bai, S.B.; Li, D.Z.; Lu, H.H.; Niu, W.Q.; Liang, W.; Bai, P.K.; Huang, Z.Q. Enhancing strength and ductility combination via tailoring the dislocation density in medium Mn steel. *Mater. Sci. Eng. A* **2023**, *879*, 145227. [CrossRef]
19. Marat, I.L.; Sunmi, S.; Bruno, C.D.C.; Seop, K.H. Micromechanical finite element analysis of strain partitioning in multiphase medium manganese TWIP+TRIP steel. *Acta Mater.* **2016**, *108*, 219–228.
20. Ding, H.; Li, H.; Misra, R.D.K.; Wu, Z.; Cai, M. Strengthening Mechanisms in Low Density Fe-26Mn-xAl-1C Steels. *Steel Res. Int.* **2017**, *89*, 1700381. [CrossRef]
21. Hu, B.; He, B.B.; Cheng, G.J.; Yen, H.W.; Luo, H.W. Super-high-strength and formable medium Mn steel manufactured by warm rolling process. *Acta Mater.* **2019**, *174*, 131–141. [CrossRef]
22. Mu, Y.K.; He, L.; Deng, S.; Jia, Y.; Jia, Y.; Wang, G.; Zhai, Q.J.; Liaw, P.K.; Liu, C.T. A high-entropy alloy with dislocation-precipitate skeleton for ultrastrength and ductility. *Acta Mater.* **2022**, *232*, 117975. [CrossRef]
23. Zhang, G.F.; Ma, W.; Ji, P.F.; Zhang, B.; Zhang, X.; Zhang, J.S.; Zhang, X.Y.; Ma, M.Z.; Liu, R.P. The effects of partial recrystallization annealing and aging treatment on the microstructural evolution and mechanical behavior of austenitic lightweight steel. *Mater Sci. Eng. A* **2023**, *885*, 145602. [CrossRef]
24. Huang, K.; Marthinsen, K.; Zhao, Q.; Roland, E.L. The double-edge effect of second-phase particles on the recrystallization behaviour and associated mechanical properties of metallic materials. *Prog. Mater. Sci.* **2018**, *92*, 284–359. [CrossRef]
25. Legros, M.; Dehm, G.; Arzt, E.; Balk, T.J. Observation of Giant Diffusivity Along Dislocation Cores. *Science* **2008**, *319*, 1646–1649. [CrossRef] [PubMed]
26. Chen, Y.A.; Li, D.Z.; Xie, R.F.; Lu, H.T. Dual precipitates and heterogeneous fine-grain structure induced strength-ductility synergy in a CoCrNi-based medium-entropy alloy. *Mater. Sci. Eng. A* **2023**, *867*, 144504. [CrossRef]
27. Humphreys, J.; Hatherly, M. *Recrystallization and Related Annealing Phenomena*, 2nd ed.; Pergamon: Bergama, Turkey, 2004.
28. Li, Z.; Wang, Y.C.; Cheng, X.W.; Li, Z.Y.; Gao, C.; Li, S.K. The effect of rolling and subsequent aging on microstructures and tensile properties of a Fe–Mn–Al–C austenitic steel. *Mater. Sci. Eng A* **2021**, *822*, 141683. [CrossRef]
29. Koyama, M.; Sawaguchi, T.; Tsuzaki, K. Effects of Si on tensile properties associated with deformation-induced ε-martensitic transformation in high Mn austenitic alloys. *Mater Trans.* **2015**, *56*, 819–825. [CrossRef]
30. Shen, S.; Xie, P.; Wu, C.; Luo, J.; Ye, H.; Chen, J. Cross-slip of extended dislocations and secondary deformation twinning in a high-Mn TWIP steel. *Int. J. Plast.* **2024**, *319*, 103922. [CrossRef]
31. Cai, W.; Sun, C.Y.; Zhang, H.J.; Wang, C.H.; Meng, L.H.; Fu, M.W. Delving into the intrinsic co-relation between microstructure and mechanical behaviour of fine-/ultrafine-grained TWIP steels via TEM and in-situ EBSD observation. *Mater. Charact.* **2024**, *210*, 113780. [CrossRef]
32. Benzing, J.T.; Poling, W.A.; Pierce, D.T.; Bentley, J.; Findley, K.O.; Raabe, D.; Wittig, J.E. Effects of strain rate on mechanical properties and deformation behavior of an austenitic Fe-25Mn-3Al-3Si TWIP-TRIP steel. *Mater. Sci. Eng. A* **2024**, *711*, 78–92. [CrossRef]
33. Boeuf, A.; Crico, S.; Roberto, G.M.; Pomot, I.; Uny, G. Effects of M23C6 precipitation on the lattice parameter of AISI 304 stainless steel. *Mater. Lett.* **1985**, *3*, 115. [CrossRef]
34. Gutierrez-Urrutia, I.; Raabe, D. Influence of Al content and precipitation state on the mechanical behavior of austenitic high-Mn low-density steel. *Scr. Mater.* **2013**, *68*, 343–347. [CrossRef]
35. Bai, S.B.; Zhao, Z.Y.; Li, D.Z.; Li, J.Y.; Bai, P.K.; Huang, Z.Q.; Lu, H.H. Realization of strength-ductility balance of 10 Mn-steel by tailoring symbiosis microstructure. *Mater. Sci. Eng. A* **2024**, *901*, 146570. [CrossRef]

materials

MDPI

*Article*

# Nonstationary Stochastic Responses of Transmission Tower-Line System with Viscoelastic Material Dampers Under Seismic Excitations

**Mingjing Chang [1], Bo Chen [1,*], Xiang Xiao [2] and Yanzhou Chen [3]**

[1] School of Civil Engineering and Architecture, Wuhan University of Technology, Wuhan 430070, China; changmj@whut.edu.cn
[2] School of Transportation and Logistics Engineering, Wuhan University of Technology, Wuhan 430070, China; xiangxiao@whut.edu.cn
[3] Central-South Architectural Design Institute Co., Ltd., Wuhan 430071, China; dly123home@163.com
* Correspondence: cebchen@whut.edu.cn

**Abstract:** The excessive vibration or collapse of a transmission tower-line (TTL) system under seismic excitation can result in significant losses. Viscoelastic material dampers (VMDs) have been recognized as an effective method for structural vibration mitigation. Most existing studies have focused solely on the dynamic analysis of TTL systems with control devices under deterministic seismic excitations. Studies focusing on the nonstationary stochastic control of TTL systems with VMDs have not been reported. To this end, this study proposes a comprehensive analytical framework for the nonstationary stochastic responses of TTL systems with VMDs under stochastic seismic excitations. The analytical model of the TTL system is formulated using the Lagrange equation. The six-parameter model of VMDs and the vibration control method are established. Following this, the pseudo-excitation method (PEM) is applied to compute the stochastic response of the controlled TTL system under nonstationary seismic excitations, and a probabilistic framework for analyzing extreme value responses is developed. A real TTL system in China is selected to verify the validity of the proposed method. The accuracy of the proposed framework is validated based on the Monte Carlo method (MCM). A detailed parametric investigation is conducted to determine the optimal damper installation scheme and examine the effects of the service temperature and site type on stochastic seismic responses. VMDs can effectively suppress the structural dynamic responses, with particularly stable control over displacement. The temperature and site type have a notable influence on the stochastic seismic responses of the TTL system. The research findings provide important references for improving the seismic performance of VMDs in TTL systems.

**Keywords:** TTL system; stochastic seismic response; VMD; control performance; nonstationary

## 1. Introduction

Earthquakes may lead to the excessive vibration and plastic failure of high-rise tower structures, resulting in structural destruction and even failure, causing significant economic losses [1,2]. Traditional methods for suppressing structural vibrations are often uneconomical because they focus on increasing stiffness to enhance its load-bearing capacity. An alternative approach to protect engineering structures is the application of various control devices [3,4]. To ensure the normal operation of transmission tower-line (TTL) systems, the application of energy dissipation devices in transmission towers is widely recognized as an effective control measure [5,6]. These energy dissipation devices can absorb and dissipate vibration energy and mitigate the dynamic impact induced by external excitations, thereby reducing the vibration amplitude of the structures, minimizing damage risk, and, ultimately, ensuring the safe operation of TTL systems [7–9]. Semi-active control devices, such as magnetorheological dampers, can effectively reduce the dynamic responses of high-rise towers. However, their configuration and control systems are commonly complicated, and the expense is relatively high. Therefore, the real application of semi-active devices in the vibration control of large truss towers has rarely been reported. To this end, various passive control devices, such as viscoelastic material dampers (VMDs), friction dampers, electromagnetic inertial mass dampers, etc., have been used in the vibration control of high-rise truss towers [10,11]. VMDs, characterized by their simple structure, easy installation, and excellent energy dissipation performance [12,13], have been widely applied in various fields, such as architecture, aerospace, and mechanical engineering, for vibration mitigation [14–16]. VMDs can suppress the vibration responses of TTL systems by increasing the overall structural damping ratio [17]. Zeng et al. [18] proposed a vibration mitigation calculation method for transmission towers with additional VMDs to determine the equivalent damping ratio. Huang et al. [19] studied the optimization design of VMDs for the response control of transmission towers.

It is noted that the majority of existing studies have focused solely on the dynamic analysis of TTL systems under deterministic seismic excitations. Seismic motions are inherently stochastic excitations with pronounced nonstationary characteristics. The seismic process, spanning from initiation to termination, is generally classified as a nonstationary stochastic phenomenon. In the seismic response control of TTL systems, the stochastic vibration analysis method based on power spectral density (PSD) is commonly taken as a robust and versatile approach. According to classical stationary stochastic vibration theory, Priestley [20] advanced the theoretical framework for assessing nonstationary stochastic responses. Most extensive research has been conducted on nonstationary stochastic seismic response analysis for high-rise and bridge structures [21–23]. Studies focusing on the nonstationary stochastic control of TTL systems with VMDs have not been reported. Zhong et al. [24] developed formulas to compute the background and resonant responses of transmission towers subjected to nonstationary downbursts by a nonstationary modal analytical method. Zhang et al. [25] developed an analytical approach to evaluate the extreme non-stationary dynamic responses of transmission towers under downbursts in the frequency domain. However, their computation focused on a single transmission tower, neglecting the coupling effects of tower-line systems.

The installation of VMDs may introduce damping non-orthogonality in a TTL system, and, thereby, the modal decomposition methods are unsuitable for solving the vibration equations. Therefore, the precise analysis of the nonstationary stochastic seismic responses of a TTL system with control devices is quite different. In this regard, a comprehensive analytical framework for the nonstationary stochastic responses of TTL systems with VMDs under stochastic seismic excitations is proposed in this study. Initially, the analytical model of the TTL system is formulated using the Lagrange equation. A six-parameter model of VMDs is introduced, and a control method based on VMDs for TTL systems is proposed. Following this, the pseudo-excitation method (PEM) is applied to compute the stochastic response of the controlled TTL system under nonstationary seismic excitations, and a probabilistic framework for analyzing extreme value responses is developed. Four damper schemes are taken into account to determine a rational VMD location. A detailed parametric investigation is conducted to investigate the effects of the service temperature and site type on stochastic seismic responses.

## 2. Model of TTL-VMD System

### *2.1. Mechanical Model of TTL*

The nonstationary stochastic dynamic computation of transmission lines requires numerous iterative calculations and is time-consuming. Therefore, the Hamilton principles and the Lagrange equation are used to develop the analytical model of transmission lines [26–29]. As displayed in Figure 1, a transmission line can be modeled as a cable by several lumped masses and elastic elements. Following the Hamilton principles, the generalized coordinates of the line can be selected as the element angle $\theta$ and the length $l$. In the in-plane direction, the kinetic energy and potential energy of the line can be determined by the Lagrange formulation. The structural equation of motion is derived using generalized coordinates:

$$\int \delta[T_c(t) - U_c(t)]\mathrm{d}t + \int \delta W_c(t)\mathrm{d}t = 0 \tag{1}$$

in which $W_c(t)$ is the virtual work of the line; $T_c$ and $U_c$ are the kinetic energy and potential energy of the line.

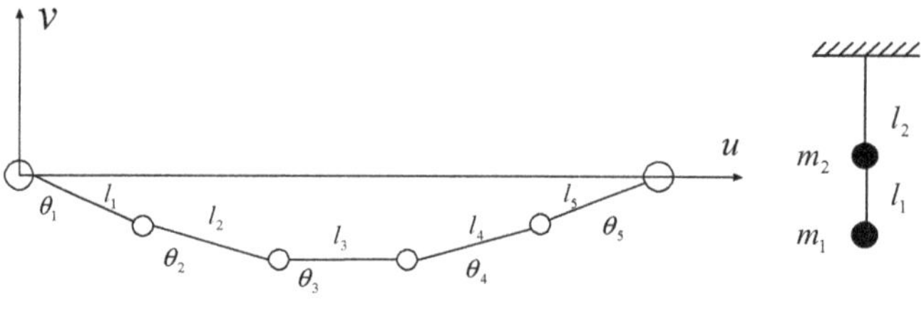


(a) (b)

**Figure 1.** Model of a transmission line for different horizontal directions: (**a**) in-plane vibration; (**b**) out-of-plane vibration.

The mass matrix $\mathbf{M}_l^{in}$ and stiffness matrix $\mathbf{K}_l^{in}$ of the line are determined by computing the partial differential of kinetic energy and potential energy to the generalized velocity and displacement. In the out-of-plane direction, the line can be simplified as a pendulum, and the structural system matrices are expressed as follows:

$$\mathbf{M}_l^{out} = \begin{bmatrix} m_1 & \\ & m_2 \end{bmatrix} \tag{2}$$

$$\mathbf{K}_l^{out} = \begin{bmatrix} \frac{m_1 g}{l_1} & -\frac{m_1 g}{l_1} \\ -\frac{m_1 g}{l_1} & \frac{m_1 g}{l_1} + \frac{(m_1+m_2)g}{l_2} \end{bmatrix} \tag{3}$$

where $\mathbf{M}_l^{out}$ and $\mathbf{K}_l^{out}$ are the structural mass and stiffness matrices, respectively.

As displayed in Figure 2a, a transmission tower is a typical high-rise flexible structure, and the analytical model can be constructed using the software ABAQUS (version 6.14). If a three-dimensional (3D) finite element model is used for real transmission towers with control devices, the nonstationary stochastic dynamic computation will be complicated and time-consuming. In this regard, lumped-mass models (Figure 2b) are commonly adopted in the stochastic dynamic analysis and vibration control of transmission towers [6]. The analytical model of the TTL is constructed as shown in Figure 3.

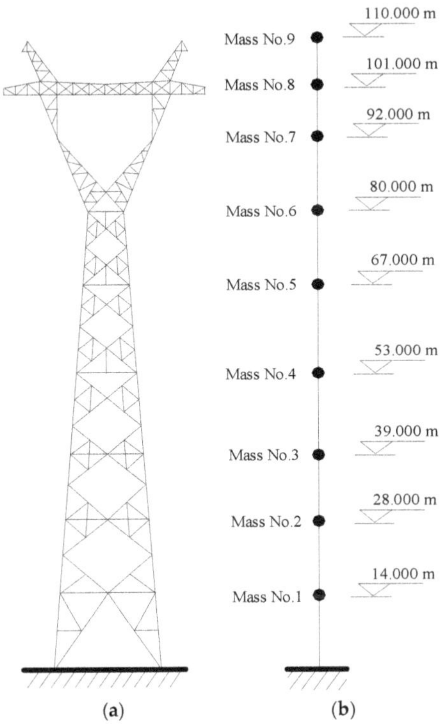


**Figure 2.** Mechanical model of the tower: (**a**) 3D finite element model; (**b**) lumped model.

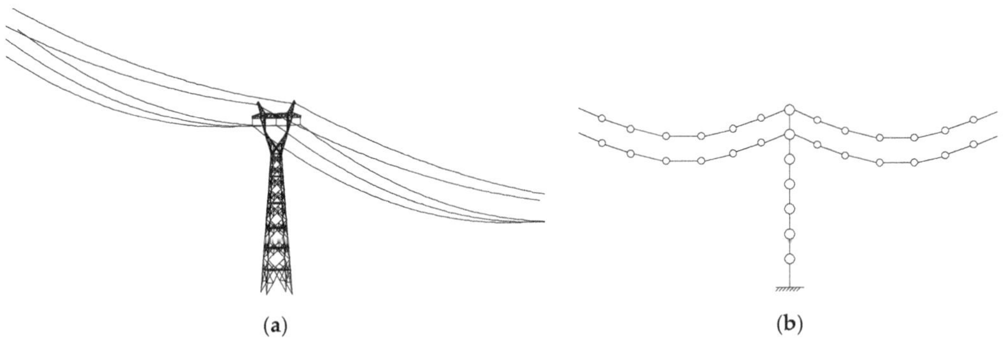


**Figure 3.** Analytical model of a TTL system: (**a**) 3D finite element model; (**b**) lumped model.

## 2.2. Mechanical Model of VMDs

Viscoelastic materials can be used in civil engineering structures to dissipate vibration energy in wind- and earthquake-induced vibration. Many factors, such as ambient temperature, frequency, and strain, may affect the behavior of viscoelastic materials. The shear stress $\tau(t)$ is given by

$$\tau(t) = G'(\omega)\gamma(t) + \frac{G''(\omega)}{\omega}\dot{\gamma}(t) \tag{4}$$

in which $\gamma(t)$ is the shear strain that varies with time $t$; $\omega$ is the frequency of the harmonic shear strain; $G'(\omega)$ is the shear storage modulus; and $G''(\omega)$ is the shear loss modulus.

To describe the frequency-dependent behavior of VMDs, many studies have employed the combination of multiple elastic springs and dashpots, such as the generalized Kelvin and generalized Maxwell models [30–33]. However, these models are relatively complex, and theoretical analysis typically requires simplicity. In this study, the six-parameter model proposed by Mazza and Vulcano is used to describe the VMD model, which can be regarded as a parallel combination of two Maxwell elements and one Kelvin element [34,35], as shown in Figure 4. The Kelvin element is composed of a dashpot with a damping coefficient $c_0$ and a spring with a stiffness coefficient $k_0$ in parallel. Each Maxwell element is composed of a dashpot with damping coefficient $c_1$ (or $c_2$) and a spring with stiffness coefficient $k_1$ (or $k_2$) connected in series.

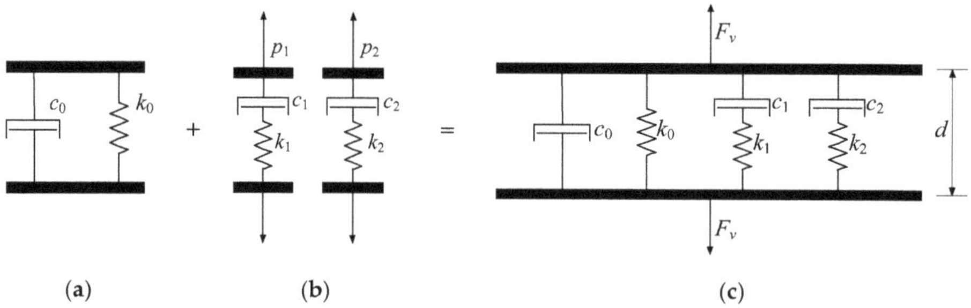


**Figure 4.** Six-parameter model of VMD: (**a**) Kelvin element; (**b**) two Maxwell elements; (**c**) six-parameter model.

The storage stiffnesses $K'$ and loss stiffness $K''$ can be expressed, for a given frequency $\omega$, as functions of the constants characterizing the behavior of the springs and dashpots constituting the six-parameter model:

$$K'(\omega) = \omega^2\left(\frac{k_1 c_1^2}{\omega^2 c_1^2 + k_1^2} + \frac{k_2 c_2^2}{\omega^2 c_2^2 + k_2^2}\right) + k_0 \tag{5}$$

$$K''(\omega) = \omega\left(\frac{k_1^2 c_1}{\omega^2 c_1^2 + k_1^2} + \frac{k_2^2 c_2}{\omega^2 c_2^2 + k_2^2} + c_0\right) \tag{6}$$

If Equations (5) and (6) are divided by the shape coefficient ($A/h$, where $A$ denotes the total shear area, and $h$ denotes the total thickness), analogous equations can be written expressing the storage modulus $G'(\omega)$ and the loss modulus $G''(\omega)$ as functions of the constants of the six elements.

Thus, the force–displacement relationship of the VMD is given by

$$F_v = k_0 d + c_0 \dot{d} + p_1 + p_2 \tag{7}$$

where $F_v$ denotes the force at both ends; $d$ and $\dot{d}$ denote the displacement and velocity of the damper, respectively; $p_1$ and $p_2$ denote the damper forces of the Maxwell elements, respectively. The relationships between the damper forces and displacement can be expressed as

$$\begin{cases} \dot{p}_1 + \mu_1 p_1 = k_1 \dot{d} \\ \dot{p}_2 + \mu_2 p_2 = k_2 \dot{d} \end{cases} \tag{8}$$

where $\mu_1$ and $\mu_2$ are the reciprocal of the relaxation time of the Maxwell elements, satisfying the conditions $\mu_1 = k_1/c_1$ and $\mu_2 = k_2/c_2$, respectively.

The damper is installed diagonally between the adjacent nodal layers in the transmission tower, as shown in Figure 5. The angle $\theta_l$ represents the angle between the damper installed in the $l$th layer ($l = 1, 2, \ldots, n$; $n$ is the number of VMDs) and the corresponding layer, as shown in Figure 5b. Since the damper force is generated by the relative displacement between the two end nodes of the damper, the damper force $F_{vl}$ in the $l$th layer can be expressed as

$$F_{vl} = k_{0l} d_l \cos\theta_l + c_{0l} \dot{d}_l \cos\theta_l + p_{1l} + p_{2l} \tag{9}$$

where $c_{0l}$ and $k_{0l}$ are the damping coefficient and stiffness coefficient of the Kelvin element for the damper in the $l$th layer, respectively; $d_l$ is the relative displacement between the $l$th layer and the $(l-1)$th layer of the transmission tower, satisfying the condition $d_l = y_l - y_{l-1}$; $y_l$ denotes the displacement of the $l$th layer; and $\theta_l$ is the angle between the damper and the $l$th layer. The damper forces of a Maxwell element are as follows:

$$\begin{cases} \dot{p}_{1l} + \mu_{1l} p_{1l} = k_{1l} \dot{d}_l \cos\theta_l \\ \dot{p}_{2l} + \mu_{2l} p_{2l} = k_{2l} \dot{d}_l \cos\theta_l \end{cases} \tag{10}$$

where $\mu_{1l}$ and $\mu_{1l}$ are the reciprocal of the relaxation time of the Maxwell elements in the $l$th layer, satisfying the conditions $\mu_{1l} = k_{1l}/c_{1l}$ and $\mu_{2l} = k_{2l}/c_{2l}$, respectively; $k_{1l}$ and $k_{2l}$ denote the stiffness coefficients of the Maxwell elements in the $l$th layer; and $c_{1l}$ and $c_{2l}$ denote the damping coefficients of the Maxwell elements in the $l$th layer.

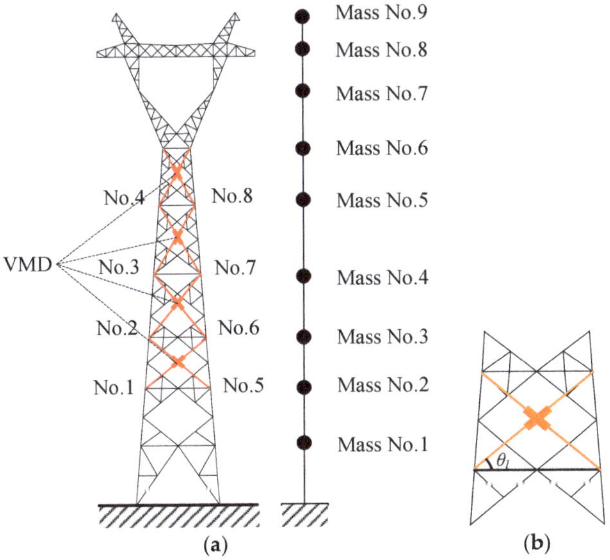


**Figure 5.** Damper installation scheme of the TTL system: (**a**) location of VMDs; (**b**) damper installation scheme.

## 3. Equation of Motion of TTL System with VMDs

To be an energy-dissipating device, VMDs can be utilized in engineering structures for vibration mitigation. The equation of motion of the controlled TTL system with VMDs under nonstationary seismic excitation is given by

$$\mathbf{M}\ddot{\mathbf{y}}(t) + \mathbf{C}\dot{\mathbf{y}}(t) + \mathbf{K}\mathbf{y}(t) + \mathbf{F}_v(t) = \mathbf{M}\mathbf{I}\ddot{x}_g(t) \tag{11}$$

where $\mathbf{y}(t)$, $\dot{\mathbf{y}}(t)$, and $\ddot{\mathbf{y}}(t)$ are the displacement, velocity, and acceleration responses of the controlled TTL system with VMDs, respectively; $\mathbf{F}_v(t)$ is the damper force vector of VMDs; **M**, **C**, and **K** are the mass, damping, and stiffness matrices of the controlled system with VMDs, respectively; **I** is the identity matrix; and $\ddot{x}_g(t)$ is the nonstationary seismic acceleration excitation, including both intensity and frequency non-stationarities.

The system matrices and vectors with VMDs are as follows:

$$\mathbf{M} = \begin{bmatrix} \mathbf{M}^{in} & 0 \\ 0 & \mathbf{M}^{out} \end{bmatrix} \tag{12}$$

$$\mathbf{K} = \begin{bmatrix} \mathbf{K}^{in} & 0 \\ 0 & \mathbf{K}^{out} \end{bmatrix} \tag{13}$$

$$\mathbf{C} = \begin{bmatrix} \mathbf{C}^{in} & 0 \\ 0 & \mathbf{C}^{out} \end{bmatrix} \tag{14}$$

$$\mathbf{y}(t) = \left[\mathbf{y}^{in}(t); \quad \mathbf{y}^{out}(t)\right] \tag{15}$$

$$\mathbf{F}_v(t) = \left[\mathbf{F}_v^{in}(t); \quad \mathbf{F}_v^{out}(t)\right] \tag{16}$$

in which $\mathbf{M}^{in}$ and $\mathbf{M}^{out}$ are the mass matrices in the two horizontal directions, respectively. Similarly, $\mathbf{K}^{in}$ and $\mathbf{K}^{out}$ are the stiffness matrices in the two horizontal directions; $\mathbf{C}^{in}$ and $\mathbf{C}^{out}$ are the Rayleigh damping matrices in the two horizontal directions; $\mathbf{y}^{in}(t)$ and $\mathbf{y}^{out}(t)$ are the system displacement in the two horizontal directions; $\mathbf{F}_v^{in}(t)$ and $\mathbf{F}_v^{out}(t)$ are the damper forces of the VMDs in the two horizontal directions.

The matrix expression of the damper force of the VMD is given by

$$\mathbf{F}_v(t) = \mathbf{\Pi}\mathbf{K}_0\mathbf{\Pi}^{\mathrm{T}}\mathbf{y}(t) + \mathbf{\Pi}\mathbf{C}_0\mathbf{\Pi}^{\mathrm{T}}\dot{\mathbf{y}}(t) + \mathbf{\Pi}\mathbf{P}_1(t) + \mathbf{\Pi}\mathbf{P}_2(t) \tag{17}$$

in which

$$\mathbf{K}_0 = \mathrm{diag}[k_{01}, k_{02}, \ldots, k_{0n}] \tag{18}$$

$$\mathbf{C}_0 = \mathrm{diag}[c_{01}, c_{02}, \ldots, c_{0n}] \tag{19}$$

$$\mathbf{\Pi} = \begin{bmatrix} \cos\theta_1 & -\cos\theta_2 & & \\ & \cos\theta_2 & \ddots & \\ & & \ddots & -\cos\theta_n \\ & & & \cos\theta_n \end{bmatrix} \tag{20}$$

where diag[·] is a diagonal matrix; $\mathbf{\Pi}$ is the position matrix of the damper forces of the VMDs.

The damper force of the Maxwell element can be expressed as:

$$\begin{cases} \dot{\mathbf{P}}_1(t) + \mathbf{U}_1\mathbf{P}_1(t) = \mathbf{K}_1\mathbf{\Pi}^{\mathrm{T}}\dot{\mathbf{y}}(t) \\ \dot{\mathbf{P}}_2(t) + \mathbf{U}_2\mathbf{P}_2(t) = \mathbf{K}_2\mathbf{\Pi}^{\mathrm{T}}\dot{\mathbf{y}}(t) \end{cases} \tag{21}$$

where

$$\begin{cases} \mathbf{P}_1(t) = [p_{11}(t), p_{12}(t), \ldots, p_{1n}(t)]^{\mathrm{T}} \\ \mathbf{P}_2(t) = [p_{21}(t), p_{22}(t), \ldots, p_{2n}(t)]^{\mathrm{T}} \end{cases} \tag{22}$$

$$\begin{cases} \mathbf{K}_1 = \mathrm{diag}[k_{11}, k_{12}, \ldots, k_{1n}] \\ \mathbf{K}_2 = \mathrm{diag}[k_{21}, k_{22}, \ldots, k_{2n}] \end{cases} \tag{23}$$

$$\begin{cases} \mathbf{U}_1 = \mathrm{diag}[\mu_{11}, \mu_{12}, \ldots, \mu_{1n}] \\ \mathbf{U}_2 = \mathrm{diag}[\mu_{21}, \mu_{22}, \ldots, \mu_{2n}] \end{cases} \tag{24}$$

# 4. Nonstationary Stochastic Seismic Responses of TTL-VMD System

## 4.1. Nonstationary Stochastic Seismic Excitation

Seismic motion is a nonstationary stochastic process, and its statistical information varies with time. It typically involves two nonstationary processes, intensity non-stationarity, and frequency non-stationarity, which can be defined using Fourier–Stieltjes integrals for nonstationary stochastic processes:

$$\ddot{x}_g(\omega,t) = \int_{-\infty}^{+\infty} a(\omega,t)e^{i\omega t}dN(\omega) \tag{25}$$

where $\ddot{x}_g(\omega,t)$ is the nonstationary stochastic seismic excitation; $i = \sqrt{-1}$; $a(\omega,t)$ denotes the deterministic modulation function for $t$ and $\omega$ and satisfies $a(\omega,t) = a^*(-\omega,t)$; "*" denotes the complex conjugate; and $N(\omega)$ is an orthogonal incremental process.

The evolutionary power spectral density (EPSD) function $S_{\ddot{x}_g}(\omega,t)$ of the nonstationary stochastic seismic excitation $\ddot{x}_g(\omega,t)$ is typically referred to as

$$S_{\ddot{x}_g}(\omega,t) = |a(\omega,t)|^2 S_{\ddot{x}_f}(\omega) \tag{26}$$

where $S_{\ddot{x}_f}(\omega)$ denotes the PSD function of a stationary process.

In most engineering calculations, the non-uniform modulation function of the non-stationary stochastic process represented by Equation (26) is approximated by a slowly varying uniform modulation function $a(t)$ to simplify the analysis [23], such that

$$S_{\ddot{x}_g}(\omega,t) = |a(t)|^2 S_{\ddot{x}_f}(\omega) \tag{27}$$

where the uniform modulation function $a(t)$ is defined as

$$a(t) = \begin{cases} (t/t_1)^2 & 0 \le t \le t_1 \\ 1 & t_1 \le t \le t_2 \\ \exp(c(t-t_2)) & t_2 \le t \end{cases} \tag{28}$$

The PSD function $S_{\ddot{x}_f}(\omega)$ is modeled using the Kanai–Tajimi spectrum [36]:

$$S_{\ddot{x}_f}(\omega) = S_0 \frac{4\xi_g^2\omega_g^2\omega^2 + \omega_g^4}{\left(\omega_g^2 - \omega^2\right)^2 + 4\xi_g^2\omega_g^2\omega^2} \tag{29}$$

where $\omega_g$ is the ground filter frequency; $\xi_g$ is the ground filter damping ratio; and $S_0$ is the spectral intensity coefficient.

## 4.2. Nonstationary Stochastic Seismic Responses

To be an efficient and accurate solution for stochastic vibration equations, the PEM can offer an effective approach for seismic response analysis of engineering structures [36]. An analytical framework based on the PEM is developed to analyze the stochastic seismic responses of the TTL-VMD system under nonstationary seismic excitation. The pseudo-excitation of the controlled TTL system $\tilde{\mathbf{F}}(\omega,t)$ is given by

$$\tilde{\mathbf{F}}(\omega,t) = \mathbf{MI}\sqrt{S_{\ddot{x}_g}(\omega,t)}e^{i\omega t} \tag{30}$$

The pseudo-response $\tilde{\mathbf{y}}(\omega,t)$ of the controlled TTL system is given by

$$\mathbf{M}\ddot{\tilde{\mathbf{y}}}(\omega,t) + \mathbf{C}\dot{\tilde{\mathbf{y}}}(\omega,t) + \mathbf{K}\tilde{\mathbf{y}}(\omega,t) + \tilde{\mathbf{F}}_v(\omega,t) = \tilde{\mathbf{F}}(\omega,t) \tag{31}$$

$$\tilde{\mathbf{F}}_v(\omega,t) = \mathbf{\Pi}\mathbf{K}_0\mathbf{\Pi}^{\mathrm{T}}\tilde{\mathbf{y}}(\omega,t) + \mathbf{\Pi}\mathbf{C}_0\mathbf{\Pi}^{\mathrm{T}}\dot{\tilde{\mathbf{y}}}(\omega,t) + \mathbf{\Pi}\tilde{\mathbf{P}}_1(\omega,t) + \mathbf{\Pi}\tilde{\mathbf{P}}_2(\omega,t) \tag{32}$$

where "~" represents the pseudo-operation symbol of the vector.

The pseudo-response $\tilde{\mathbf{y}}(\omega,t)$ can be obtained by performing a time-history response analysis at each discrete frequency point $\omega_j$ ($j = 1, 2, \ldots, M$); $M$ is the number of frequency points with equal intervals. Given the initial responses of the TTL-VMD system, the equation of motion is rewritten in the state space. Thus, Equations (31) and (32) can be transformed into

$$\dot{\mathbf{v}}(\omega_j,t) = \mathbf{H}\mathbf{v}(\omega_j,t) + \mathbf{r}(\omega_j,t) \tag{33}$$

For the original TTL system without control, the system matrices are

$$\mathbf{v}(\omega_j,t) = \begin{bmatrix} \tilde{\mathbf{y}}(\omega_j,t) \\ \dot{\tilde{\mathbf{y}}}(\omega_j,t) \end{bmatrix};\ \mathbf{r}(\omega_j,t) = \begin{bmatrix} 0 \\ \mathbf{M}^{-1}\tilde{\mathbf{F}}(\omega_j,t) \end{bmatrix} \tag{34}$$

$$\mathbf{H} = \begin{bmatrix} 0 & \mathbf{I} \\ -\mathbf{M}^{-1}\mathbf{K} & -\mathbf{M}^{-1}\mathbf{C} \end{bmatrix} \tag{35}$$

For the TTL system controlled by VMDs, the system matrices are

$$\mathbf{v}(\omega_j,t) = \begin{bmatrix} \tilde{\mathbf{y}}(\omega_j,t) \\ \dot{\tilde{\mathbf{y}}}(\omega_j,t) \\ \tilde{\mathbf{P}}_1(\omega_j,t) \\ \tilde{\mathbf{P}}_2(\omega_j,t) \end{bmatrix};\ \mathbf{r}(\omega_j,t) = \begin{bmatrix} 0 \\ \mathbf{M}^{-1}\tilde{\mathbf{F}}(\omega_j,t) \\ 0 \\ 0 \end{bmatrix} \tag{36}$$

$$\mathbf{H} = \begin{bmatrix} 0 & \mathbf{I} & 0 & 0 \\ -\mathbf{M}^{-1}(\mathbf{K} + \mathbf{\Pi}\mathbf{K}_0\mathbf{\Pi}^{\mathrm{T}}) & -\mathbf{M}^{-1}(\mathbf{C} + \mathbf{\Pi}\mathbf{C}_0\mathbf{\Pi}^{\mathrm{T}}) & -\mathbf{M}^{-1}\mathbf{\Pi} & -\mathbf{M}^{-1}\mathbf{\Pi} \\ 0 & \mathbf{K}_1\mathbf{\Pi}^{\mathrm{T}} & -\mathbf{U}_1 & 0 \\ 0 & \mathbf{K}_2\mathbf{\Pi}^{\mathrm{T}} & 0 & -\mathbf{U}_2 \end{bmatrix} \tag{37}$$

To further improve the solution accuracy, a numerical integration algorithm, namely, the precise integration method, is employed for the response calculation of the TTL-VMD system [37]. According to the precise integration method, the load variation within each time step $\mathbf{r}(\omega_j,t)$ is linear:

$$\mathbf{r}(\omega_j,t) = \mathbf{r}_0(\omega_j,t) + \mathbf{r}_1(\omega_j,t) \times (t - t_0) \tag{38}$$

where $\mathbf{r}_0(\omega_j,t)$ and $\mathbf{r}_1(\omega_j,t)$ are parameter vectors determined by $\mathbf{r}(\omega_j,t)$, respectively.

Based on the precise integration method, the state vector $\mathbf{v}(\omega_j,t_m)$ at time $t_m$ has the following iterative relationship:

$$\begin{aligned} \mathbf{v}(\omega_j,t_m) = \mathbf{T}(\tau)(\mathbf{v}(\omega_j,t_{m-1}) + \mathbf{H}(\mathbf{r}_0(\omega_j,t_{m-1}) + \mathbf{H}^{-1}\mathbf{r}_1(\omega_j,t_{m-1}))) \\ -\mathbf{H}(\mathbf{r}_0(\omega_j,t_m) + \mathbf{H}^{-1}\mathbf{r}_1(\omega_j,t_m) + \mathbf{r}_1(\omega_j,t_m)\tau) \end{aligned} \tag{39}$$

$$\tau = t_{m+1} - t_m \tag{40}$$

where $\mathbf{T}(\tau)$ is determined by the precise integration method.

The EPSD matrix of the nonstationary stochastic responses of the TTL-VMD system is

$$\mathbf{S}_{\mathbf{vv}}(\omega_j,t_m) = \mathbf{v}(\omega_j,t_m)^{*}\mathbf{v}(\omega_j,t_m)^{\mathrm{T}} \tag{41}$$

The response variance of the controlled system is given by

$$\sigma_{\mathbf{v}}^2(t_m) = 2\int_0^{+\infty} \mathbf{S_{vv}}(\omega, t_m)\mathrm{d}\omega \tag{42}$$

*4.3. Extreme Responses of Controlled System*

The extreme responses of structures are of utmost concern in engineering design and safety assessment. Thus, extreme responses can be employed in the stochastic seismic control of a TTL system to evaluate the efficacy of the VMDs. To derive the extreme responses, the nonstationary stochastic responses of the TTL-VMD system can be replaced by the equivalent stationary stochastic responses over a specific duration $\tau$. Assuming $y(t)$ is any nonstationary response, its equivalent stationary response is represented by $\overline{y}(t)$. Let the extreme value of the equivalent stationary response over the duration $\tau$ be $\overline{y}_e$, with a standard deviation of $\sigma_{\overline{y}}$, and define the dimensionless parameter as $\eta = \overline{y}_e/\sigma_{\overline{y}}$. Based on the assumption that the number of threshold crossings follows a Poisson process, the probability distribution of $\eta$ is given by

$$\mathrm{P}(\eta) = \exp\left[-\kappa\tau\exp\left(-\frac{\eta^2}{2}\right)\right] \tag{43}$$

$$\kappa = \frac{1}{\pi}\sqrt{\frac{\lambda_2}{\lambda_0}};\ \lambda_0 = 2\int_0^{+\infty} S_{\overline{yy}}(\omega)\mathrm{d}\omega;\ \lambda_2 = 2\int_0^{+\infty} \omega^2 S_{\overline{yy}}(\omega)\mathrm{d}\omega \tag{44}$$

where $\lambda_0$ is the 0th-order spectral moment, which reflects the displacement response variance; $\lambda_2$ is the 2nd-order spectral moment, which reflects the acceleration response variance; $S_{\overline{yy}}(\omega)$ is the equivalent stationary PSD over the duration $\tau$; and $\tau$ is defined as the duration during which the intensity exceeds 50% of the peak vibration:

$$S_{\overline{yy}}(\omega) = \frac{1}{\tau}\int_{t_0/\sqrt{2}}^{t_0/\sqrt{2}+\tau} S_{yy}(\omega, t)\mathrm{d}t \tag{45}$$

where $S_{yy}$ is any diagonal element of the EPSD matrix $\mathbf{S_{vv}}$.

The expected value of $\eta$ and the extreme response $\overline{y}_e$ are approximately determined based on the following probability distribution:

$$\mathrm{E}(\eta) = \sqrt{2\ln(\kappa\tau)} + \frac{\gamma}{\sqrt{2\ln(\kappa\tau)}} \tag{46}$$

$$\overline{y}_\mathrm{e} = \mathrm{E}(\eta)\sigma_{\overline{y}} = \left(\sqrt{2\ln(\kappa\tau)} + \frac{\gamma}{\sqrt{2\ln(\kappa\tau)}}\right)\sigma_{\overline{y}} \tag{47}$$

where $\gamma$ = 0.5772 is the Euler constant.

To assess the efficacy of VMDs under nonstationary excitation, the reduction rate ($R$) of the extreme responses is defined as

$$R = \frac{\overline{y}_e^o - \overline{y}_e^c}{\overline{y}_e^o} \tag{48}$$

where $\overline{y}_e^o$ and $\overline{y}_e^c$ are the extreme responses of the original system and the controlled system, respectively.

## 5. Case Study

### *5.1. Analytical Parameters*

A TTL system in Figure 2 is adopted to verify the validity of the proposed approach using VMDs under nonstationary stochastic seismic excitation. The case study involves a transmission tower with a height of 110 m and a span of 832 m. The cross-sectional area of the transmission line is 71.25 mm$^2$, and the density is 1.43 kg/m. The first two natural frequencies of the tower in the in-plane direction are 0.57 Hz and 2.29 Hz, while the counterparts in the out-of-plane direction are 0.65 Hz and 2.25 Hz. The damping ratios of the first two frequencies are set as 0.01.

The parameters for the modulation function of the nonstationary stochastic seismic evolutionary spectrum in Equation (27) are determined based on the ground acceleration power spectrum specified by the Chinese code [38], namely, $t_1$ = 0.8 s, $t_2$ = 7 s, and $c$ = 0.35. The site type of the TTL system is II according to the Chinese code. The parameters for the stationary PSD in Equation (29) are as follows: $\omega_g$ = 12.57 rad/s, $\xi_g$ = 0.72, and $S_0$ = 250.57 cm$^2$/s$^3$. The mean of the maximum ground acceleration corresponding to this seismic parameter type is 0.4 g (where g is the gravitational acceleration), which is the earthquake intensity of grade 8 specified by the Chinese code. The precise integration method adopts a time step of 0.01 s. To ensure the convergence of the integrals, the trapezoidal rule is adopted in the numerical integration. The TTL system has a relatively low natural frequency, and the power spectrum decays rapidly at high frequencies, ensuring that the contribution from high-frequency components is negligible. Thus, the cutoff frequency is selected as 60 rad/s. The frequency step is selected as 0.1 rad/s to maintain stability and ensure convergence.

### *5.2. Damper Installation Schemes*

Overall, eight VMDs are installed in the tower body. Four dampers are located in the in-plane direction and the other four are located in the out-of-plane direction. The parameters of the VMDs under different service temperatures are listed in Table 1. In reality, the six-parameter model of VMDs allows, unlike a single Kelvin model or Maxwell model, a better description of the variation in the storage modulus $G'(\omega)$ and the loss modulus $G''(\omega)$ for varying values of the excitation frequency. Indeed, a single Kelvin model or Maxwell model can be calibrated to match exactly the experimental values of $G'(\omega)$ and $G''(\omega)$ corresponding only to a single value of the frequency at a given temperature. The parameters are determined by fitting experimental data using the least squares approximation under different temperature conditions [34,35].

**Table 1.** Parameters of the six-parameter model under different temperatures.

| $T$ (°C) | $k_0$ (kN/cm) | $k_1$ (kN/cm) | $k_2$ (kN/cm) | $c_0$ (kN·s/cm) | $c_1$ (kN·s/cm) | $c_2$ (kN·s/cm) |
|---|---|---|---|---|---|---|
| 21 | 0.36 | 42.08 | 6.87 | 0.37 | 0.83 | 2.15 |
| 32 | 0.21 | 23.54 | 3.18 | 0 | 0.39 | 0.73 |
| 38 | 0.28 | 2327.46 | 3.28 | 0 | 0.57 | 0.18 |

To investigate the control efficacy of VMDs, four damper installation schemes are taken into consideration. For Scheme No. 1, all eight VMDs are installed at the tower bottom. Four VMDs are on the first floor and the other four dampers are on the second floor. For Scheme No. 2, all the VMDs are evenly installed from the first to the fourth floor with two dampers on each floor. For Scheme No. 3, all the VMDs are evenly installed from the third to the sixth floor with two dampers on each floor, as shown in Figure 5a. For Scheme No. 4, all eight VMDs are installed on top of the tower body, namely, four VMDs on the fifth floor and the other four on the sixth floor.

Figure 6 displays the extreme responses of the tower at an exceeding probability of 0.85 for different control schemes. The labels of the vertical axis in the figure (Mass) represent the number of the concentrated mass layers in the lumped mass model as displayed in Figure 2b, namely, the number of structural dynamic degrees of freedom. The extreme responses of the tower are suppressed substantially with the placement of VMDs. The control effectiveness of the extreme displacement and the extreme velocity are much better than that of the extreme acceleration. In addition, the control effectiveness of the extreme displacement significantly surpasses that of the extreme velocity. The overall efficacy of Schemes No. 3 and 4 is substantially better than that of Schemes No. 1 an2. The efficacy of Scheme No. 2 is slightly superior to that of Scheme No. 1. Scheme No. 4 exhibits the best control effectiveness in both directions in comparison with the other schemes. Regarding Scheme No. 4, the displacement responses on top of the tower body are much larger than those of the tower bottom. Relatively large floor drifts and damper deformation can be observed. Thus, the energy dissipation and the control efficacy of Scheme No. 4 are satisfactory in comparison with the other three schemes.

The reduction rates of the extreme response of different control schemes are computed and listed in Table 2. The displacement reduction rates of the tower top of Scheme No. 4 are 31.90% and 36.92%, respectively, which are slightly better than the counterparts of Scheme No. 3. Specifically, the reduction rates of the extreme displacement responses in both Schemes No. 3 and 4 exceed 30%, while the reduction rates of extreme velocity responses are also approximately 30%. The overall control performance of Schemes No. 1 and 2 is unsatisfactory because the reduction rates are no more than 20%. The control efficacy of velocity responses is slightly inferior to that of displacement responses, while the reduction rate of acceleration responses is the smallest one. It is noted that structural safety is assessed based on displacement and internal forces. Thus, the reduction rate of displacement responses is the most important index in control performance. In summary, all four damper installation schemes can mitigate the structural seismic responses in both directions. It is noted that the above results were obtained at 21 °C, and similar observations can be made at other service temperatures. A comprehensive evaluation reveals that Scheme No. 4 exhibits a significant advantage in overall control performance, leading to the decision to adopt the damper arrangement of Scheme No. 4 at 21 °C for subsequent parametric studies. The proposed algorithm framework was tested on a computer with the configuration of Intel Core i7-9700 CPU @ 3.00 GHz and 32.0 GB RAM. The calculation times for the uncontrolled and controlled TTL systems are 53 s and 68 s, respectively.

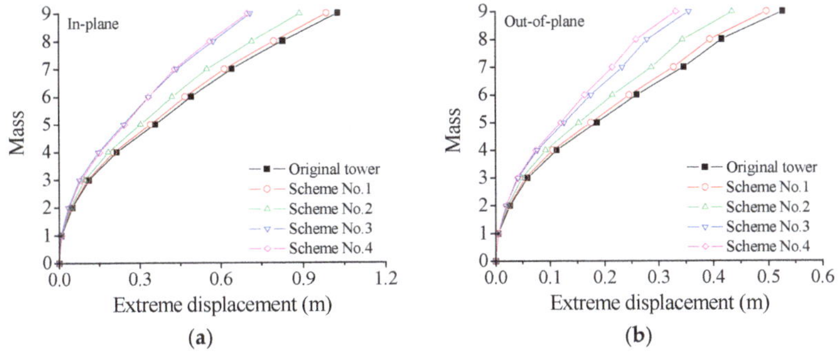


**Figure 6.** *Cont.*

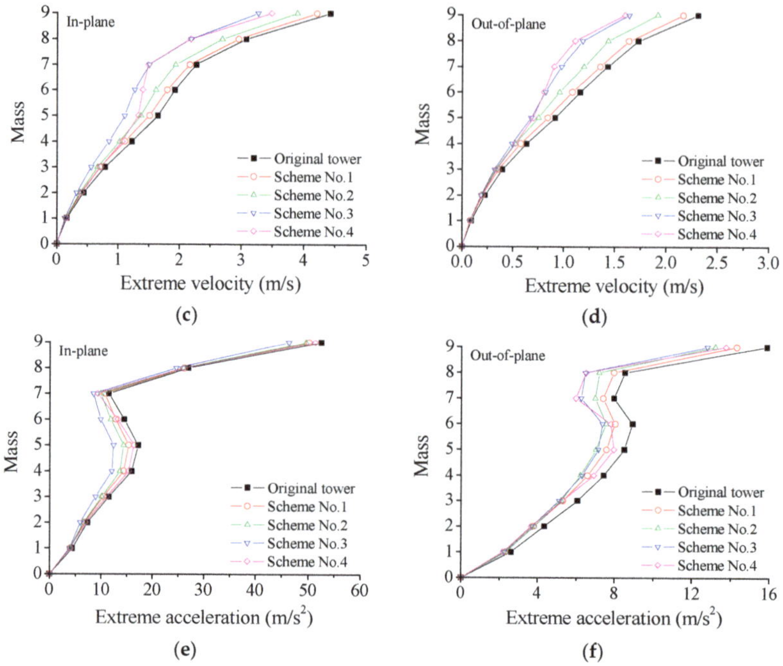


**Figure 6.** Control efficacy of different schemes: (**a**) extreme displacement; (**b**) extreme displacement; (**c**) extreme velocity; (**d**) extreme velocity; (**e**) extreme acceleration; and (**f**) extreme acceleration.

**Table 2.** Reduction rates of the extreme response of different control schemes.

| Direction | Location | Response | Reduction Rate (%) | | | |
|---|---|---|---|---|---|---|
| | | | Scheme No. 1 | Scheme No. 2 | Scheme No. 3 | Scheme No. 4 |
| In-plane | Mass No. 6 (top of tower body) | Displacement | 4.71 | 14.71 | 32.57 | 32.49 |
| | | Velocity | 6.55 | 16.29 | 34.08 | 27.25 |
| | | Acceleration | 11.51 | 18.47 | 31.92 | 8.76 |
| | Mass No. 8 (cross arm) | Displacement | 3.86 | 13.30 | 30.53 | 32.17 |
| | | Velocity | 3.98 | 12.62 | 29.00 | 29.14 |
| | | Acceleration | 2.75 | 3.66 | 8.95 | 3.85 |
| | Mass No. 9 (Tower top) | Displacement | 4.08 | 13.54 | 30.85 | 31.90 |
| | | Velocity | 4.97 | 12.02 | 26.08 | 21.17 |
| | | Acceleration | 4.59 | 5.69 | 11.89 | 2.35 |
| Out-of-plane | Mass No. 6 (top of tower body) | Displacement | 5.63 | 17.44 | 32.56 | 37.06 |
| | | Velocity | 6.62 | 17.27 | 28.99 | 30.25 |
| | | Acceleration | 10.06 | 15.33 | 17.56 | 12.64 |
| | Mass No. 8 (cross arm) | Displacement | 5.24 | 17.26 | 32.96 | 37.97 |
| | | Velocity | 5.30 | 16.93 | 31.67 | 35.89 |
| | | Acceleration | 6.57 | 15.72 | 24.40 | 23.84 |
| | Mass No. 9 (tower top) | Displacement | 5.54 | 17.33 | 32.41 | 36.92 |
| | | Velocity | 6.32 | 17.13 | 29.23 | 30.85 |
| | | Acceleration | 9.93 | 16.81 | 19.37 | 13.39 |

### *5.3. Control Performance*

The Monte Carlo method (MCM) was employed to evaluate the accuracy of the proposed nonstationary seismic response analysis framework for the TTL system with VMDs. A numerical simulation was conducted using 5000 samples, followed by statistical analysis of the response RMS at different time instances. The time-varying displacement RMS of the nonstationary responses of the TTL system was computed by PEM in comparison with those based on MCM, as displayed in Figure 7a,b. The observations made with both methods exhibit excellent agreement, thereby validating the accuracy of the computational approach proposed in this study. Similar observations are seen for the velocity and acceleration responses. The response RMS of the TTL system initially increases rapidly and then decreases slowly. In addition, it is seen that the in-plane response RMS is larger than that in the out-of-plane direction for all three types of dynamic responses. This is because the in-plane stiffness of the TTL system is much larger than the out-of-plane stiffness, which is the action of transmission lines. Furthermore, the response RMS of the tower top (Mass No. 9) is much larger than that of the tower body (Mass No. 6), which means that obvious whiplash effects can be observed.

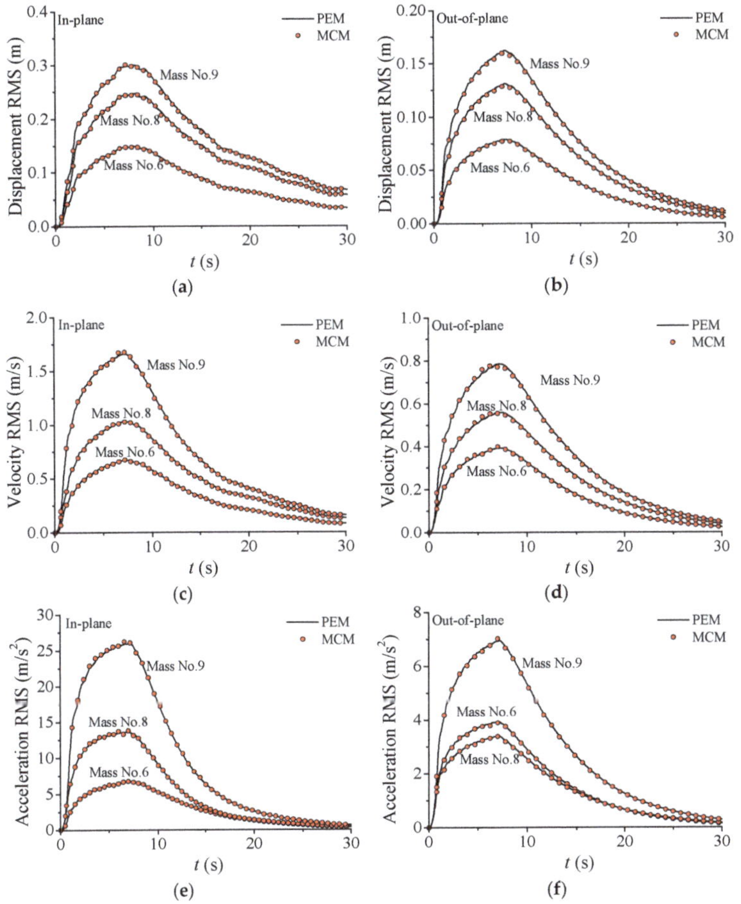


**Figure 7.** Nonstationary stochastic seismic responses of TTL system: (**a**) displacement RMS; (**b**) displacement RMS; (**c**) velocity RMS; (**d**) velocity RMS; (**e**) acceleration RMS; (**f**) acceleration RMS.

The EPSD of the nonstationary seismic responses of the TTL system without and with VMDs for the in-plane vibration are computed and displayed in Figure 8. The comparison of EPSD without and with control indicates that there is no significant change in the structural natural frequencies, suggesting that VMDs have a minimal influence on the inherent vibrational characteristics of the TTL system. It is noted that the spectral amplitude of the controlled tower shows a marked decrease in comparison with that of the original tower. The out-of-plane EPSD presents similar trends, as displayed in Figure 9.

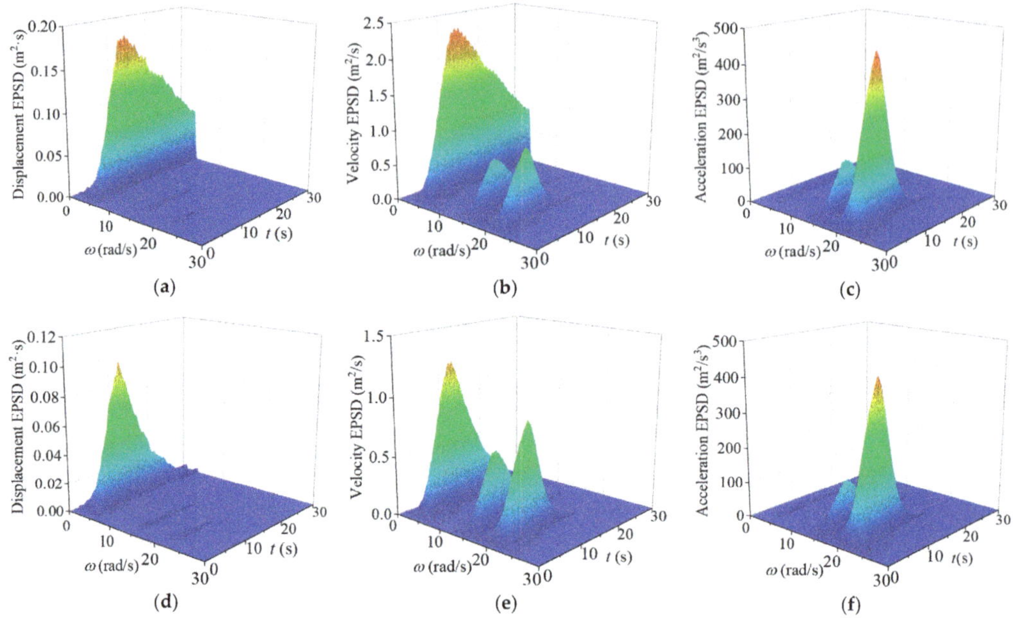


**Figure 8.** EPSD of structural responses for in-plane vibration: (**a**) displacement EPSD (original tower); (**b**) velocity EPSD (original tower); (**c**) acceleration EPSD (original tower); (**d**) displacement EPSD (with control); (**e**) velocity EPSD (with control); (**f**) acceleration EPSD (with control).

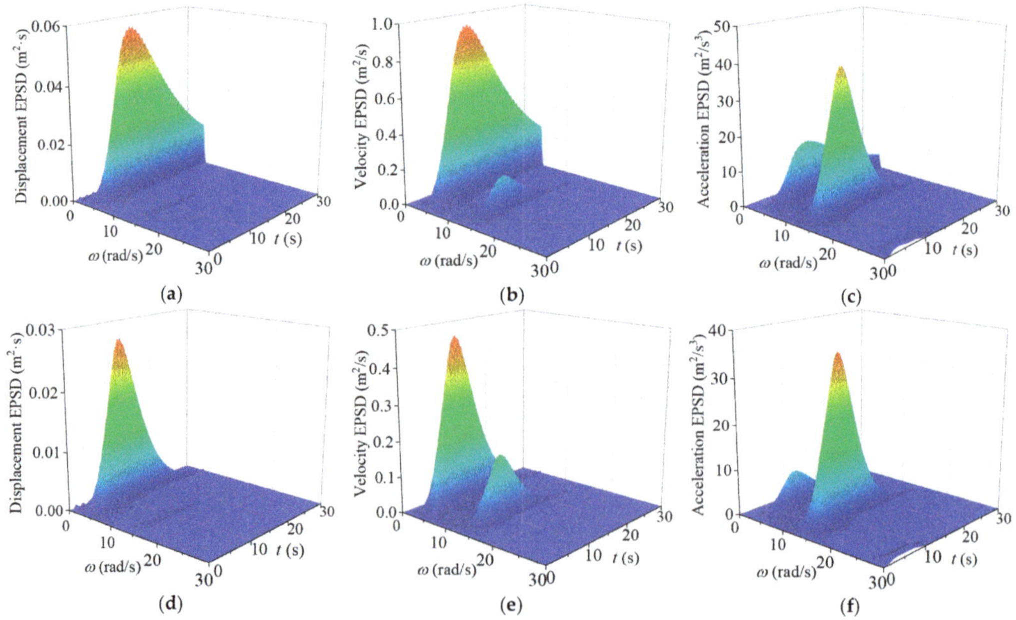


**Figure 9.** EPSD of structural responses for out-of-plane vibration: (**a**) displacement EPSD (original tower); (**b**) velocity EPSD (original tower); (**c**) acceleration EPSD (original tower); (**d**) displacement EPSD (with control); (**e**) velocity EPSD (with control); (**f**) acceleration EPSD (with control).

Table 3 lists the extreme values of the EPSD and the reduction rates. Notably, the spectral extreme value of the dynamic responses in both directions shows a remarkable decrease. The data indicate that the reduction rate of the spectrum at the first-order frequency reaches approximately 50%, while the reduction rate at the second-order frequency is significantly lower than that at the first-order frequency. In addition, the out-of-plane reduction rates of the spectra are slightly larger than those for the in-plane vibration, which is similar to the observations of the extreme responses in Table 2. The application of VMDs demonstrates satisfactory control performance in reducing displacement and velocity responses but is less effective in reducing the acceleration response.

**Table 3.** Extreme values of EPSD and reduction rates.

| Direction | Response | Freq. Order | Extreme Value | | Reduction Rate (%) |
|---|---|---|---|---|---|
| | | | Original Tower | With Control | |
| In-plane | Displacement ($m^2 \cdot s$) | 1 | 0.1833 | 0.1002 | 45.33 |
| | | 2 | 0.0035 | 0.0030 | 13.17 |
| | Velocity ($m^2/s$) | 1 | 2.3491 | 1.2569 | 46.50 |
| | | 2 | 0.7019 | 0.6199 | 11.69 |
| | Acceleration ($m^2/s^3$) | 1 | 30.5683 | 16.8119 | 45.00 |
| | | 2 | 147.1099 | 130.4146 | 11.35 |
| Out-of-plane | Displacement ($m^2 \cdot s$) | 1 | 0.0581 | 0.0280 | 51.71 |
| | | 2 | 0.0011 | 0.0010 | 12.28 |
| | Velocity ($m^2/s$) | 1 | 0.9676 | 0.4815 | 50.24 |
| | | 2 | 0.2074 | 0.1841 | 11.24 |
| | Acceleration ($m^2/s^3$) | 1 | 16.2968 | 8.3393 | 48.83 |
| | | 2 | 40.9186 | 36.5927 | 10.57 |

## 6. Parametric Study

### *6.1. Effect of Service Temperature*

It is reported that the performance of VMDs may vary with the temperatures to some extent [30,31]. Thus, a detailed analysis is conducted on the nonstationary extreme responses of TTL systems under three service temperatures, as shown in Table 1. Figure 10 presents the extreme seismic responses of the tower at an exceeding probability of 0.85 under different service temperatures. The curves indicate that the service temperature significantly affects the VMDs' performance. The extreme responses gradually reduce with the increasing temperature. Notably, at 21 °C, the extreme seismic responses are smallest compared with those at 32 °C and 38 °C. The effects of temperature on displacement and velocity are much larger than those on acceleration.

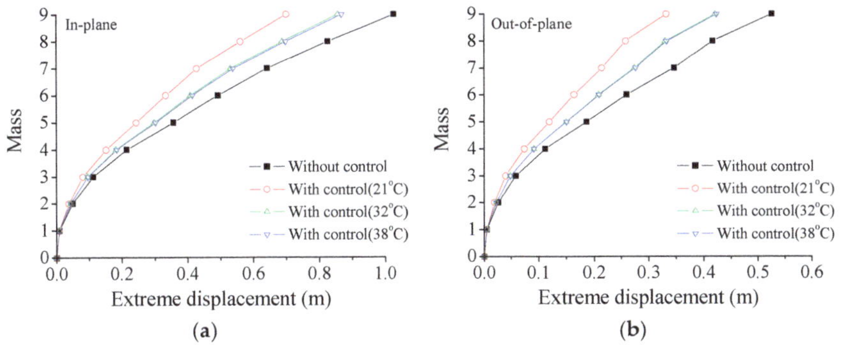


**Figure 10.** *Cont.*

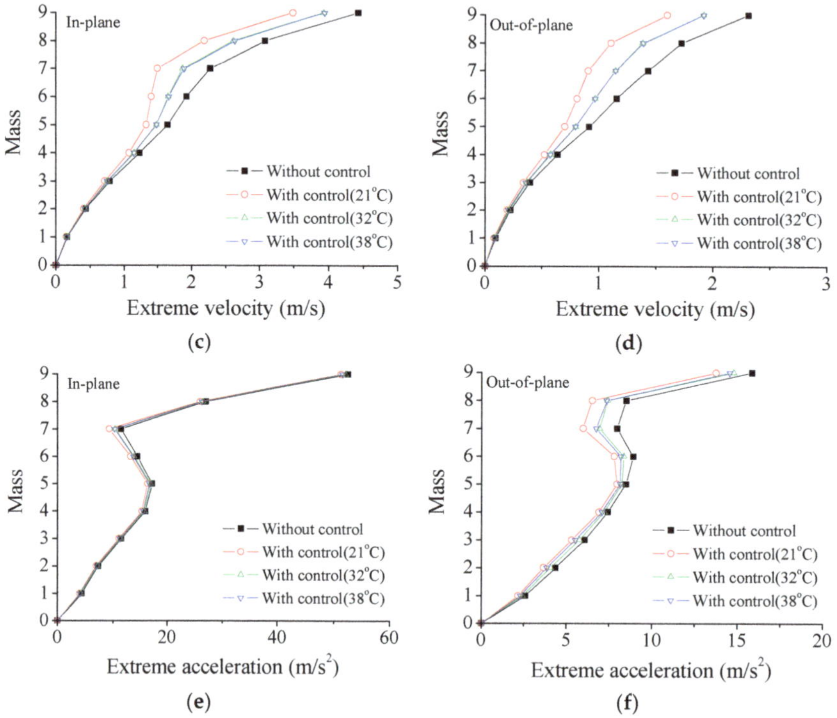


**Figure 10.** Extreme responses of the tower under different service temperatures: (**a**) extreme displacement; (**b**) extreme displacement; (**c**) extreme velocity; (**d**) extreme velocity; (**e**) extreme acceleration; (**f**) extreme acceleration.

The reduction rates of extreme responses at different locations are computed and compared in Table 4 to quantify the influence of the service temperature on the control efficacy. The reduction rates of the extreme displacement at 21 °C are about 32% and 37% for the in-plane and out-of-plane vibration, respectively. The counterparts of the extreme velocity responses at 21 °C are about 21% and 31%, respectively. However, the reduction rates of the extreme responses at 32 °C and 38 °C are substantially lower than that at 21 °C. In addition, the reduction rates of the acceleration responses are much smaller than those of the displacement and velocity responses. Furthermore, the displacement reduction rates surpass those of the velocity and acceleration, confirming that the control performance of the displacement responses is the most effective one. It is noted that the control performance of the VMDs remarkably decreases with the increasing service temperature. This phenomenon can be attributed to the variations in the rheological properties of the viscoelastic material, which result in a decrease in the energy dissipation efficiency of VMDs as the temperature increases. Therefore, it is possible to adopt advanced viscoelastic materials designed to ensure more stable performance of VMDs over a wide thermal range or explore thermal regulation strategies.

**Table 4.** Reduction rates of the extreme response at different service temperatures.

| Direction | Location | Response | Reduction Rate (%) | | |
|---|---|---|---|---|---|
| | | | 21 °C | 32 °C | 38 °C |
| In-plane | Mass No. 6 (top of tower body) | Displacement | 32.49 | 16.96 | 15.73 |
| | | Velocity | 27.25 | 14.24 | 13.71 |
| | | Acceleration | 8.76 | 4.03 | 5.24 |
| | Mass No. 8 (cross arm) | Displacement | 32.17 | 17.45 | 16.19 |
| | | Velocity | 29.14 | 17.89 | 16.85 |
| | | Acceleration | 3.85 | 9.72 | 10.73 |
| | Mass No. 9 (tower top) | Displacement | 31.90 | 16.67 | 15.44 |
| | | Velocity | 21.17 | 11.20 | 10.93 |
| | | Acceleration | 2.35 | 0.87 | 1.74 |
| Out-of-plane | Mass No. 6 (top of tower body) | Displacement | 37.06 | 20.12 | 19.58 |
| | | Velocity | 30.25 | 16.57 | 16.68 |
| | | Acceleration | 12.64 | 6.22 | 8.22 |
| | Mass No. 8 (cross arm) | Displacement | 37.97 | 20.81 | 20.29 |
| | | Velocity | 35.89 | 20.03 | 19.91 |
| | | Acceleration | 23.84 | 13.06 | 15.56 |
| | Mass No. 9 (tower top) | Displacement | 36.92 | 20.08 | 19.56 |
| | | Velocity | 30.85 | 16.95 | 16.99 |
| | | Acceleration | 13.39 | 6.86 | 8.36 |

### 6.2. *Effect of Site Type*

The seismic excitations acting on the upper structures are tightly related to the properties of the site type, such as the ground filter frequency, ground filter damping ratio, and the spectral intensity coefficient, as displayed in Equation (29). Therefore, the effects of the site type on the control performance of the TTL system with VMDs are investigated in this section. Listed in Table 5 are the seismic parameters for different site types under earthquake intensity 8. It is noted that the soil becomes softer with increasing site type.

**Table 5.** Parameters of different site types (earthquake intensity 8).

| Site Type | $\omega_g$ (Rad/s) | $\xi_g$ | $S_0$ (cm$^2$/s$^3$) |
|---|---|---|---|
| I | 15.71 | 0.64 | 215.19 |
| II | 12.57 | 0.72 | 250.57 |
| III | 8.98 | 0.80 | 323.85 |
| IV | 6.61 | 0.90 | 398.29 |

The variations in the extreme responses with the site types are computed and displayed in Figure 11. Data analysis reveals that as soil conditions become softer, both extreme displacement and velocity exhibit a significant increasing trend, while extreme acceleration shows a decreasing trend. This indicates that the site type has a notable influence on the stochastic dynamic responses. It is observed that the VMDs exhibit satisfactory control efficacy for all four site types. The reduction rates of extreme responses with different site types are analyzed and listed in Table 6. It is seen that the reduction rate of extreme displacement exceeds 30%. The reduction rate of extreme velocity is comparatively lower than that of extreme displacement, averaging around 25%. The reduction rate of extreme acceleration is much smaller than the counterparts of extreme displacement and velocity. It is seen that the reduction rates of extreme velocity and acceleration present an upward trend with the increase in the site type number. However, with the increase in the site type number, the soil layer softens, and the reduction rate of extreme displacement stabilizes, indicating a stable control performance on displacement responses.

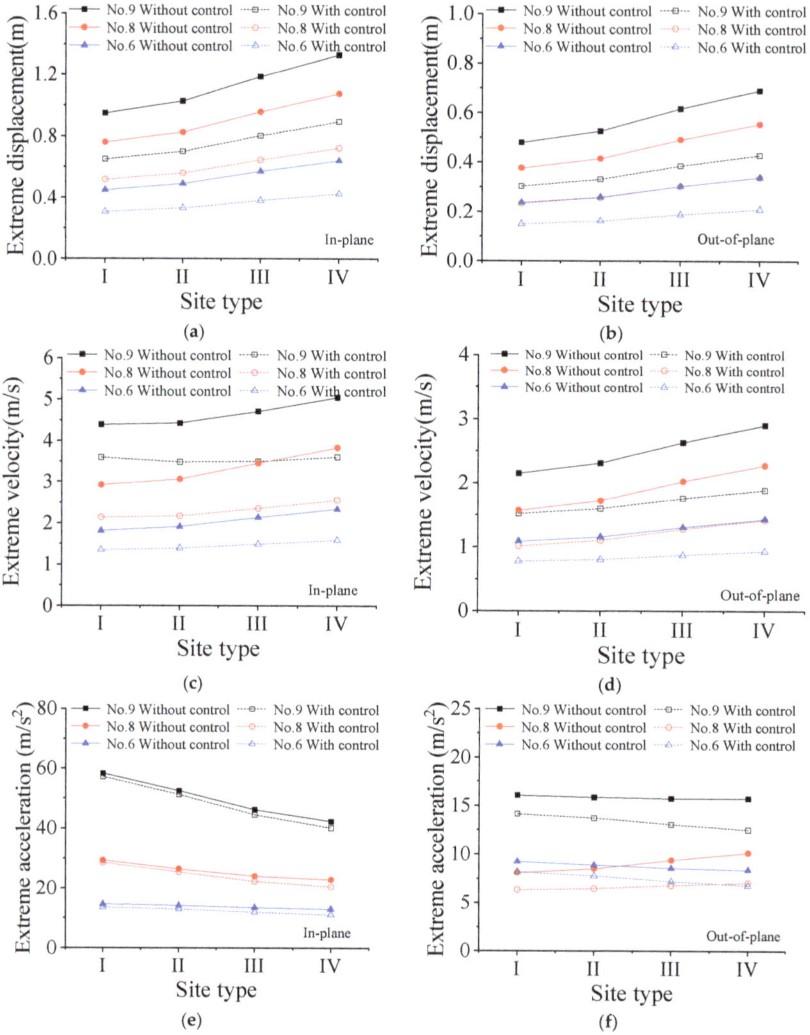


**Figure 11.** Extreme responses of the tower with different site types: (**a**) extreme displacement; (**b**) extreme displacement; (**c**) extreme velocity; (**d**) extreme velocity; (**e**) extreme acceleration; (**f**) extreme acceleration.

**Table 6.** Reduction rates of the extreme responses with different site types.

| Direction | Location | Response | Reduction Rate (%) | | | |
|---|---|---|---|---|---|---|
| | | | I | II | III | IV |
| In-plane | Mass No. 6 (top of tower body) | Displacement | 32.17 | 32.49 | 32.98 | 33.39 |
| | | Velocity | 25.43 | 27.25 | 30.06 | 32.04 |
| | | Acceleration | 7.34 | 8.76 | 11.16 | 14.26 |
| | Mass No. 8 (cross arm) | Displacement | 31.94 | 32.17 | 32.56 | 32.87 |
| | | Velocity | 26.83 | 29.14 | 31.78 | 33.23 |
| | | Acceleration | 2.60 | 3.85 | 7.08 | 10.34 |
| | Mass No. 9 (tower top) | Displacement | 31.50 | 31.90 | 32.41 | 32.73 |
| | | Velocity | 18.02 | 21.17 | 25.63 | 28.56 |
| | | Acceleration | 1.83 | 2.35 | 3.60 | 5.13 |

**Table 6.** *Cont.*

| Direction | Location | Response | Reduction Rate (%) | | | |
|---|---|---|---|---|---|---|
| | | | I | II | III | IV |
| Out-of-plane | Mass No. 6 (top of tower body) | Displacement | 36.83 | 37.06 | 37.54 | 37.92 |
| | | Velocity | 28.45 | 30.25 | 32.98 | 34.76 |
| | | Acceleration | 11.49 | 12.64 | 15.75 | 18.99 |
| | Mass No. 8 (cross arm) | Displacement | 37.82 | 37.97 | 38.32 | 38.65 |
| | | Velocity | 35.42 | 35.89 | 36.77 | 37.49 |
| | | Acceleration | 21.72 | 23.84 | 27.60 | 30.39 |
| | Mass No. 9 (tower top) | Displacement | 36.72 | 36.92 | 37.39 | 37.81 |
| | | Velocity | 29.41 | 30.85 | 33.18 | 34.81 |
| | | Acceleration | 12.00 | 13.39 | 16.94 | 20.46 |

## 7. Conclusions

This paper proposes a comprehensive analytical framework for the nonstationary stochastic responses of TTL systems with VMDs under stochastic seismic excitations. A detailed parametric investigation was carried out to determine the optimal damper installation scheme and examine the effects of the service temperature and site type on stochastic seismic responses. The conclusions drawn from the case studies are as follows:

(1) The nonstationary stochastic responses of the TTL system are consistent with those based on MCM, validating the accuracy of the proposed analytical framework and demonstrating the effectiveness of VMDs in stochastic response reduction of the TTL system. The VMDs exhibit satisfactory control performance in reducing displacement, while the influence on acceleration is relatively minor.

(2) All four damper installation schemes can mitigate the seismic responses of the TTL system. Scheme No. 4 exhibits the best control effectiveness in both directions in comparison with the other schemes.

(3) Temperature significantly influences the control performance of VMDs, with optimal control effects observed at 21 °C. In contrast, higher temperatures result in a relatively lower reduction rate of stochastic dynamic responses.

(4) The site type has a notable influence on the stochastic seismic responses of the TTL system. Under different site conditions, VMD is effective in controlling the extreme responses of transmission towers, with particularly stable control over displacement. As soil conditions become softer, both extreme displacement and velocity exhibit a significant increasing trend, while extreme acceleration shows a decreasing trend.

(5) It is noted that the proposed comprehensive analytical framework is versatile and applicable to the seismic response analysis of other types of TTL systems with VMDs.

**Author Contributions:** Conceptualization, B.C.; methodology, B.C. and X.X.; software, M.C.; validation, B.C. and M.C.; formal analysis, X.X. and Y.C.; investigation, B.C. and Y.C.; resources, Y.C. and X.X.; data curation, Y.C.; writing—original draft preparation, B.C. and M.C.; writing—review and editing, B.C. and M.C.; visualization, B.C. and M.C.; supervision, B.C. and M.C.; project administration, B.C.; funding acquisition, B.C. and Y.C. All authors have read and agreed to the published version of the manuscript.

**Funding:** This research was funded by the National Natural Science Foundation of China (52278528) and the Science and Technology Plan Project of the Central-South Architectural Design Institute Co., Ltd. (2021-KY-067).

**Institutional Review Board Statement:** Not applicable.

**Informed Consent Statement:** Not applicable.

**Data Availability Statement:** The data generated and/or analyzed during this study are not publicly available due to privacy and our ongoing research.

**Acknowledgments:** The authors sincerely appreciate the selfless support provided by Wuhan University of Technology and Central-South Architectural Design Institute Co., Ltd. during the progress of this study.

**Conflicts of Interest:** Author Yanzhou Chen was employed by the company Central-South Architectural Design Institute Co., Ltd. The remaining authors declare that the research was conducted in the absence of any commercial or financial relationships that could be construed as a potential conflict of interest.

## Abbreviations

| **Symbol** | **Description** |
|---|---|
| $a(t)$ | Uniform modulation function for $t$ |
| $a(\omega,t)$ | Modulation function for $t$ and $\omega$ |
| $A, h$ | Total shear area/total thickness |
| $c_0, c_1, c_2$ | Damping coefficients of the six-parameter model |
| $c_{0l}, c_{1l}, c_{2l}$ | Damping coefficients of the six-parameter model in the $l$th layer |
| $\mathbf{C}$ | Damping matrix of the transmission tower-line system |
| $\mathbf{C}_0$ | Damping matrix composed of all the parameters $c_0$ in the VMD system |
| $\mathbf{C}^{in}$, $\mathbf{C}^{out}$ | Damping matrix in the in-plane/out-of-plane direction |
| $d$; $\dot{d}$ | Displacement/velocity of the damper |
| $d_l$ | Relative displacement between the $l$th layer and the $(l-1)$th layer of the transmission tower |
| $F_v$ | Damper force |
| $\mathbf{F}_v$ | Damper force vector |
| $\mathbf{F}_v^{in}(t)$, $\mathbf{F}_v^{out}(t)$ | Damper force vector of the VMDs in the in-plane/out-of-plane direction |
| $\tilde{\mathbf{F}}(\omega,t)$ | Pseudo-excitation vector of the system |
| $\tilde{\mathbf{F}}_v(\omega,t)$ | Pseudo-damper force vector of the system |
| $g$ | Gravitational acceleration |
| $G'(\omega)$, $G''(\omega)$ | Shear storage/loss modulus |
| $\mathbf{H}$ | State matrix in the equation of state |
| $\mathbf{I}$ | Identity matrix |
| $k_0, k_1, k_2$ | Stiffness coefficients of the six-parameter model |
| $k_{0l}, k_{1l}, k_{2l}$ | Stiffness coefficients of the six-parameter model in the $l$th layer |
| $K', K''$ | Storage/loss stiffness |
| $\mathbf{K}$ | Stiffness matrix of the transmission tower-line system |
| $\mathbf{K}_0, \mathbf{K}_1, \mathbf{K}_2$ | Stiffness matrix composed of all the parameters $k_0/k_1/k_2$ in the VMD system |
| $\mathbf{K}_l^{in}$, $\mathbf{K}_l^{out}$ | Stiffness matrix of the transmission line in the in-plane/out-of-plane direction |
| $\mathbf{K}^{in}$, $\mathbf{K}^{out}$ | Stiffness matrix in the in-plane/out-of-plane direction |
| $M$ | Number of frequency points with equal intervals |
| $\mathbf{M}$ | Mass matrix of the transmission tower-line system |
| $\mathbf{M}_l^{in}$, $\mathbf{M}_l^{out}$ | Mass matrix of the transmission line in the in-plane/out-of-plane direction |
| $\mathbf{M}^{in}$, $\mathbf{M}^{out}$ | Mass matrix of the transmission tower-line system in the in-plane/out-of-plane direction |
| $n$ | Damper number in the tower |
| $N(\omega)$ | Orthogonal incremental process |
| $p_1, p_2$ | Damper forces of the Maxwell elements |
| $\mathbf{P}_1(t)$, $\mathbf{P}_2(t)$ | Damper force vectors of the Maxwell element |
| $\tilde{\mathbf{P}}_1(\omega,t)$, $\tilde{\mathbf{P}}_2(\omega,t)$ | Pseudo-damper force vectors of the Maxwell element |
| $\mathbf{r}(\omega,t)$ | Excitation vector in the equation of state |
| $\mathbf{r}_0(\omega,t)$, $\mathbf{r}_1(\omega,t)$ | Parameter vectors determined by $\mathbf{r}(\omega,t)$ |

| | |
|---|---|
| $R$ | Reduction rate of the extreme response |
| $S_0$ | Spectral intensity coefficient |
| $S_{\ddot{x}_f}(\omega)$ | Power spectral density function of a stationary process |
| $S_{\ddot{x}_g}(\omega, t)$ | Evolutionary power spectral density function of the nonstationary stochastic seismic excitation |
| $S_{\overline{y}\overline{y}}(\omega)$ | Equivalent stationary power spectral density function |
| $\mathbf{S}_{\mathbf{vv}}(\omega, t)$ | Power spectral density function of the state vector in the equation of state |
| $T_c$, $U_c$ | Kinetic/potential energy of the transmission line |
| $t$ | Time variable |
| $\mathbf{U}_1$, $\mathbf{U}_2$ | Matrix composed of the reciprocal $\mu_1/\mu_2$ of the relaxation time of the Maxwell elements |
| $\mathbf{v}(\omega, t)$ | Pseudo-state vector in the equation of state |
| $W_c(t)$ | Virtual work of the transmission line |
| $\ddot{x}_g(t)$ | Nonstationary seismic acceleration excitation |
| $y(t)$ | Nonstationary response |
| $y_l$ | Displacement of the transmission tower-line system in the $l$th layer |
| $\overline{y}(t)$ | Equivalent stationary response |
| $\overline{y}_e$ | Extreme value of the equivalent stationary response |
| $\overline{y}_e^o$, $\overline{y}_e^c$ | Extreme responses of the original/controlled system |
| $\mathbf{y}(t)$, $\dot{\mathbf{y}}(t)$, $\ddot{\mathbf{y}}(t)$ | Displacement/velocity/acceleration responses vector |
| $\mathbf{y}^{in}(t)$, $\mathbf{y}^{out}(t)$ | Displacement in the in-plane/out-of-plane direction |
| $\tilde{\mathbf{y}}(\omega, t)$, $\dot{\tilde{\mathbf{y}}}(\omega, t)$ | Pseudo-displacement/velocity |
| $\tau(t)$, $\gamma(t)$ | Shear stress/strain |
| $\sigma_{\mathbf{v}}^2(t_m)$ | Response variance of the system at time $t_m$ |
| $\omega$ | Circular frequency |
| $\omega_g$ | Ground filter frequency |
| $\xi_g$ | Ground filter damping ratio |
| $\mu_1$, $\mu_2$ | Reciprocal of the relaxation time of the Maxwell elements |
| $\mu_{1l}$, $\mu_{1l}$ | Reciprocal of the relaxation time of the Maxwell elements in the $l$th layer |
| $\theta_l$ | Angle between the damper installed in the $l$th layer |
| $\mathbf{\Pi}$ | Position matrix of the damper forces of the VMDs |
| $\lambda_0$, $\lambda_2$ | 0th-order/2nd-order spectral moment |
| $\tau$ | Duration during which the intensity exceeds 50% of the peak vibration |
| $\eta$ | Ratio of extreme value and standard deviation of the equivalent stationary response |
| diag[·] | Diagonal matrix |
| P(·) | Probability distribution |
| **Acronym** | **Description** |
| EPSD | Evolutionary power spectral density |
| MCM | Monte Carlo Method |
| PEM | Pseudo-excitation method |
| PSD | Power spectral density |
| RMS | Root mean square |
| TTL | Transmission tower-line |
| TTL-VMD | Transmission tower-line and viscoelastic material damper |
| VMD | Viscoelastic material damper |
| 3D | Three-dimensional |

## References


1. Tian, L.; Ma, R.S.; Li, H.N.; Wang, Y. Progressive collapse of power transmission tower-line system under extremely strong earthquake excitations. *Int. J. Struct. Stab. Dyn.* **2016**, *16*, 1550030. [CrossRef]
2. Chen, B.; Guo, W.H.; Li, P.Y.; Xie, W.P. Dynamic responses and vibration control of the transmission tower-line system: A state-of-the-art review. *Sci. World J.* **2014**, *2014*, 538457. [CrossRef] [PubMed]

3. Kim, D.W.; Kwak, M.K.; Kim, S.M.; Feeny, B.F. Slewing and active vibration control of a flexible single-link manipulator. *Actuators* **2025**, *14*, 43. [CrossRef]
4. Hasheminejad, S.M.; Lissek, H.; Vesal, R. Energy harvesting and inter-floor impact noise control using an optimally tuned hybrid damping system. *Acta Acust.* **2024**, *8*, 42. [CrossRef]
5. Xie, Q.; Sun, L. Failure mechanism and retrofitting strategy of transmission tower structures under ice load. *J. Constr. Steel Res.* **2012**, *74*, 26–36. [CrossRef]
6. Chen, B.; Weng, S.; Zhi, L.H.; Li, D.M. Response control of a large transmission tower line system under seismic excitations using friction dampers. *Adv. Struct. Eng.* **2017**, *20*, 1155–1173. [CrossRef]
7. Chen, B.; Sun, Y.Z.; Li, Y.L.; Zhao, S.L. Control of seismic response of a building frame by using hybrid system with magnetorheological dampers and isolators. *Adv. Struct. Eng.* **2014**, *17*, 1199–1215. [CrossRef]
8. Versaci, M.; Palumbo, A. Magnetorheological fluids: Qualitative comparison between a mixture model in the Extended Irreversible Thermodynamics framework and a Herschel–Bulkley experimental elastoviscoplastic model. *Int. J. Non-Linear Mech.* **2020**, *118*, 103288. [CrossRef]
9. Versaci, M.; Angiulli, G. A FEMs magnetic-thermal study for a MR automotive damper. In Proceedings of the Photonics & Electromagnetics Research Symposium (PIERS), Hangzhou, China, 21–25 November 2021. [CrossRef]
10. Chen, B.; Zheng, J.; Qu, W.L. Control of wind-induced response of transmission tower-line system by using magnetorheological dampers. *Int. J. Struct. Stab. Dyn.* **2009**, *9*, 661–685. [CrossRef]
11. Chen, B.; Yang, D.; Feng, K.; Ouyang, Y.Q. Response control of a high-rise television tower under seismic excitations by friction dampers. *Int. J. Struct. Stab. Dyn.* **2018**, *18*, 1850140. [CrossRef]
12. Palmeri, A.; Ricciardelli, F.; De Luca, A.; Muscolino, G. State space formulation for linear viscoelastic dynamic systems with memory. *J. Eng. Mech. ASCE* **2003**, *129*, 715–724. [CrossRef]
13. Palmeri, A.; Muscolino, G. A numerical method for the time-domain dynamic analysis of buildings equipped with viscoelastic dampers. *Struct. Control Health Monit.* **2011**, *18*, 519–539. [CrossRef]
14. Tsai, C.S.; Lee, H.H. Applications of viscoelastic dampers to high-rise buildings. *J. Struct. Eng. ASCE* **1993**, *119*, 1222–1233. [CrossRef]
15. Shu, Z.; You, R.; Zhou, Y. Viscoelastic materials for structural dampers: A review. *Constr. Build. Mater.* **2022**, *342*, 127955. [CrossRef]
16. Chang, T.S.; Singh, M.P. Seismic analysis of structures with a fractional derivative model of viscoelastic dampers. *Earthq. Eng. Eng. Vib.* **2002**, *2*, 251–260. [CrossRef]
17. Gan, F.L.; Jiang, H.L. Wind-induced vibration control for transmission system using steel-lead viscoelastic damper. *Adv. Mater. Dev. Appl. Mech.* **2014**, *597*, 376–379. [CrossRef]
18. Zeng, C.; Hao, D.X.; Hou, L.Q. Research on calculation of equivalent damping ratio of electrical transmission tower line system with viscoelastic dampers. *Adv. Mater. Res.* **2014**, *1064*, 115–119. [CrossRef]
19. Huang, G.P.; Hu, J.H.; He, Y.Z.; Liu, H.B.; Sun, X.G. Optimization of VEDs for vibration control of transmission line tower. *Adv. Civ. Eng.* **2021**, *2021*, 9060414. [CrossRef]
20. Priestley, M.B. Evolutionary spectra and nonstationary processes. *J. R. Stat. Soc.* **1965**, *27*, 204–237. [CrossRef]
21. Zhang, Y.H.; Li, Q.S.; Lin, J.H.; Williams, F. W Random vibration analysis of long-span structures subjected to spatially varying ground motions. *Soil Dyn. Earthq. Eng.* **2009**, *29*, 620–629. [CrossRef]
22. Jia, H.Y.; Zhang, D.Y.; Zheng, S.X.; Xie, W.C.; Pandey, M.D. Local site effects on a high-pier railway bridge under tridirectional spatial excitations: Nonstationary stochastic analysis. *Soil Dyn. Earthq. Eng.* **2013**, *52*, 55–69. [CrossRef]
23. Zhao, Y.; Si, L.T.; Ouyang, H. Frequency domain analysis method of nonstationary random vibration based on evolutionary spectral representation. *Eng. Comput.* **2018**, *35*, 1098–1127. [CrossRef]
24. Zhong, Y.L.; Li, S.; Jin, W.C.; Yan, Z.T.; Liu, X.P.; Li, Y. Frequency domain analysis of along-wind response and study of wind loads for transmission tower subjected to downbursts. *Buildings* **2022**, *12*, 148. [CrossRef]
25. Zhang, Y.; Wang, D.; Sun, C.; Fu, X. Probabilistic study on non-stationary extreme response of transmission tower under moving downburst impact. *Structures* **2024**, *68*, 107179. [CrossRef]
26. Li, H.N.; Shi, W.L.; Wang, G.X.; Jia, L.G. Simplified models and experimental verification for coupled transmission tower–line system to seismic excitations. *J. Sound Vib.* **2005**, *286*, 569–585. [CrossRef]
27. Jamalabadi, M.Y.A. An improvement of port-hamiltonian model of fluid sloshing coupled by structure motion. *Water* **2018**, *10*, 1721. [CrossRef]
28. Chen, B.; Song, X.X.; Li, W.P.; Wu, J.B. Vibration control of a wind-excited transmission tower-line system by shape memory alloy dampers. *Materials* **2022**, *15*, 1790. [CrossRef]
29. Song, X.X.; Chen, Y.Z.; Chen, B.; Wu, J.B. Response control of a transmission tower-line system under wind excitations by electromagnetic inertial mass dampers. *Adv. Struct. Eng.* **2022**, *25*, 3334–3348. [CrossRef]
30. Makris, N. Complex-parameter kelvin model for elastic foundations. *Earthq. Eng. Struct. Dyn.* **1994**, *23*, 251–264. [CrossRef]

31. Singh, M.P.; Verma, N.P.; Moreschi, L.M. Seismic analysis and design with Maxwell dampers. *J. Eng. Mech. ASCE* **2003**, *129*, 273–282. [CrossRef]
32. Singh, M.P.; Chang, T.S.; Nandan, H. Algorithms for seismic analysis of MDOF systems with fractional derivatives. *Eng. Struct.* **2011**, *33*, 2371–2381. [CrossRef]
33. Renaud, F.; Dion, J.L.; Chevallier, G.; Tawfiq, I.; Lemaire, R. A new identification method of viscoelastic behavior: Application to the generalized Maxwell model. *Mech. Syst. Signal Process.* **2011**, *25*, 991–1010. [CrossRef]
34. Mazza, F.; Vulcano, A. Control of the along-wind response of steel framed buildings by using viscoelastic or friction dampers. *Wind Struct.* **2007**, *10*, 233–247. [CrossRef]
35. Mazza, F.; Vulcano, A. Control of the earthquake and wind dynamic response of steel-framed buildings by using additional braces and/or viscoelastic dampers. *Earthq. Eng. Struct. Dyn.* **2011**, *40*, 155–174. [CrossRef]
36. Zhao, Y.; Li, Y.Y.; Zhang, Y.H.; Kennedy, D. Nonstationary seismic response analysis of long-span structures by frequency domain method considering wave passage effect. *Soil Dyn. Earthq. Eng.* **2018**, *109*, 1–9. [CrossRef]
37. Song, G.; Lin, J.H.; Williams, F.W.; Wu, Z.G. Precise integration strategy for aseismic LQG control of structures. *Int. J. Numer. Methods Eng.* **2006**, *68*, 1281–1300. [CrossRef]
38. *GB 50011-2010*; China Ministry of Construction. Code for Seismic Design of Buildings. Chinese Architectural Industry Press: Beijing, China, 2016. (In Chinese)

*materials*

MDPI

*Article*

# Deformation Behavior of S32750 Duplex Stainless Steel Based on In Situ EBSD Technology

**Shun Bao, Han Feng *, Zhigang Song, Jianguo He, Xiaohan Wu and Yang Gu**

Central Iron & Steel Research Institute Co., Ltd., Beijing 100081, China; baoshun1994@163.com (S.B.); zhigangsongnercast@163.com (Z.S.); hejianguo@nercast.com (J.H.); wuxiaohan@nercast.com (X.W.); guyang@nercast.com (Y.G.)
* Correspondence: fenghannercast@163.com

**Abstract:** In this study, we investigated the two-phase hardening behavior and microstructural evolution of S32750 duplex stainless steel during the tensile deformation process. The analysis was conducted using in situ electron backscatter diffraction (EBSD), scanning electron microscopy (SEM), and microhardness testing. It was observed that strain transfer occurred between the two phases in the position away from the fracture. The ferrite phase exhibited softening, while the austenite phase underwent hardening. In the region less than 1 mm from the fracture site, both phases experienced a rapid hardening, with the maximum hardness difference between the two phases near the fracture reaching approximately 45 HV. In situ EBSD results indicate that the kernel average misorientation (KAM) value for the ferrite phase consistently exceeds that of the austenite phase during the initial stages of deformation. Conversely, in the final stages of deformation, the KAM value for austenite surpasses that of ferrite. In the initial stage of deformation, the type of grain boundaries in both phases remains largely unaltered. However, in the later stages of deformation, there is a marked increase in the number of small-angle grain boundaries within ferrite, which become approximately three times that of the large-angle grain boundaries. As deformation progresses, the maximum orientation distribution density of the ferrite phase is reduced by approximately 50%, with the preferred orientation shifting from the {100} plane to the {111} plane. In contrast, the orientation distribution of the austenite remains relatively uniform, with no significant change in the maximum orientation distribution density observed. This indicates that after substantial deformation, the rotation of ferrite grains significantly increases the deformation resistance, whereas the austenite phase continues to harden. This differential behavior leads to the continuous accumulation of strain at the phase boundaries, ultimately causing cracks to form at these boundaries and resulting in the sample's fracture.

**Keywords:** S32750 duplex stainless steel; in situ EBSD; deformation mechanism; orientation distribution

## 1. Introduction

The S32750 duplex stainless steel (DSS) matrix comprises two phases: ferrite and austenite. Through appropriate solution treatment, harmful phases can be eliminated and also ensure that the ferrite and austenite contents are nearly equal [1]. S32750 duplex stainless steel, renowned for its superior mechanical properties and outstanding corrosion resistance, is extensively utilized in the petrochemical and marine engineering industries [2–4]. After proper forging and hot rolling processes, the two phases in duplex stainless steel—ferrite and austenite—are typically uniformly distributed throughout the sheets. Due to the distinct structures of these phases, their deformation capabilities and behaviors vary significantly [5–7]. During the deformation process, the two phases in duplex stainless steel must not only withstand deformation together but also coordinate their responses, indicating that the underlying deformation mechanism is inherently more complex. Fu et al. [8] conducted a study on the deformation behavior of the austenitic phase in duplex stainless steel under ultra-high-cycle fatigue conditions. They found that hardening and softening phenomena are induced by the proliferation of dislocations and the plastic transfer, which in turn triggers the initiation of microcracks at the phase boundaries. The uneven distribution and transfer of load between the two phases are identified as the primary contributors to the onset of cracking. Zhang et al. [9] subjected the surface of S32750 duplex stainless steel to significant deformation using the ultrasonic surface rolling process. This treatment introduced a multitude of small-angle grain boundaries and compressive residual stress. Consequently, these alterations facilitated the enrichment of $Cr_2O_3$ during the corrosion process, thereby enhancing the steel's overall corrosion resistance. Mariana et al. investigated the structural integrity of S32750 duplex stainless steel under thermal deformation conditions. They discovered that when the deformation exceeded 20% at 1050 °C, cracks tended to form on the side of the material. In contrast, no side cracking was observed when the deformation reached 50% at 1300 °C. This difference was primarily attributed to the formation and subsequent dissolution of the σ phase [10]. Zhao et al. [11] observed that under high-temperature conditions, a low strain rate resulted in dynamic recovery of duplex stainless steel. Conversely, at a lower temperature of 900 °C, an increase in strain rate corresponded to a higher likelihood of dynamic recrystallization occurring. During thermal deformation, temperature emerges as the predominant factor influencing the recrystallization of austenite. Notably, at high temperatures, the recrystallization behavior is largely independent of the strain rate [12]. Currently, numerous studies have been conducted on the deformation behavior of duplex stainless steel at both room temperature and high temperature. However, the relationship between the deformation behavior of the two phases and the failure behavior of the material remains underexplored.

The deformation coordination and behavior of duplex stainless steel during the deformation process play a crucial role in determining the material's strength, plasticity, and fracture mechanism. Currently, research on the deformation behavior of duplex stainless steel predominantly concentrates on the effects of hot deformation characteristics, dynamic recovery, recrystallization, and the cold deformation process on its properties. However, there remains a significant gap in systematic and comprehensive studies regarding the two-phase hardening laws of duplex stainless steel throughout the entire deformation process, the evolution of its microstructure, and the impact of these changes on the material's susceptibility to cracking. The interplay between the deformation behaviors of the two phases is intricately linked to the steel's failure mechanisms. In this study, we utilized in situ EBSD technology to investigate the deformation behaviors of S32750 duplex stainless steel throughout the entire tensile process. The objective of this research is to elucidate the fracture mechanism of the steel and offer insights for enhancing its mechanical properties.

## 2. Materials and Methods

The chemical composition of S32750 DSS hot rolled plate is as follows (in wt.%): 0.023% C, 0.31% Si, 0.67% Mn, 25.51% Cr, 6.34% Ni, 3.92% Mo, 0.25% N, and the balance Fe. To achieve a uniform two-phase structure and eliminate the precipitated phases, the test steel was subjected to solution treatment at 1080 °C, held for 40 min, and subsequently quenched in water. The mechanical property test was conducted in strict accordance with the standard GB/T 228.1-2021. The test was performed at room temperature, and the tensile rate was maintained at 0.00025 $s^{-1}$. The testing equipment used was a GNT100

Universal Testing Machine (It is produced in Beijing, China and independently developed by Iron & Steel Research Institute Nake Testing Technology Co., Ltd.). The mechanical properties of the hot rolled plate were yield strength 570 MPa, tensile strength 816 MPa, elongation 41%, and shrinkage 68%. The in situ EBSD experiments encompassed several key components: real-time stretching using the HS4000 apparatus, scanning observation with the Thermo Scientific Apreo 2C (It is produced in the United States and held by Beijing Zhongke Kefu Technology Co., Ltd. of Beijing, China) scanning electron microscope (SEM), and acquisition of two-phase crystallology information via the EDAX Velocity Super detector. The dimensions and sampling orientation of the in situ tensile samples are illustrated in Figure 1. The applied tensile strain rate for these experiments was 2 μm/s. The microhardness test was performed using the Japanese FM-300 digital microhardness tester (It is produced by FUTURE-TECH in Japan and provided by Kunshan Fuze Testing Equipment Co., Ltd., Kunshan, China), and the FEI Quanta 650 FEG SEM (Made in the United States, provided by Beijing Yuanhaiwei Technology Co., Ltd., Beijing, China) was used to observe the fracture morphology of the sample. The grain boundaries of the two phases under different states were statistically analyzed using Aztec Crystal software (Version number: 2.1).

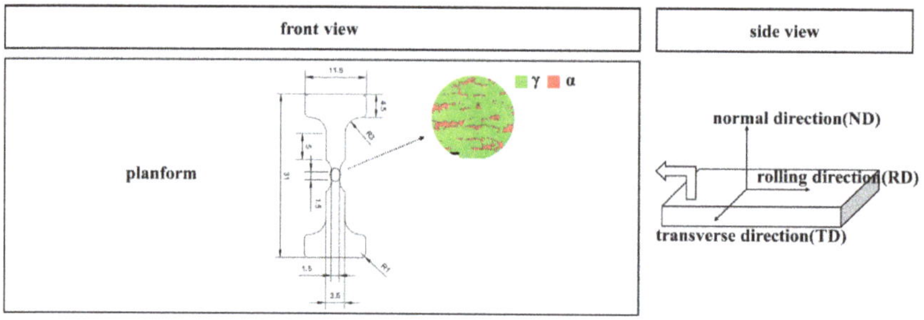


**Figure 1.** In situ tensile specimen size and sampling location.

The matrix of the test steel consists of ferrite and austenite. After hot rolling and solution treatment, the two phases are uniformly distributed throughout the matrix. Different etching methods reveal different morphologies. After electrolytic etching with potassium hydroxide, the two phases in the substrate are distinctly revealed, as shown in Figure 2a, where the black areas represent the ferrite phase and the white areas represent the austenite phase. The volume fractions of ferrite and austenite were 48.2% and 51.8%, respectively. After being soaked and corroded by a mixture of KOH and $H_2SO_4$, the grain boundaries of the two phases become evident, as shown in Figure 2b. The grain sizes of the two phases were calculated using Nano Measure software (version 1.2.0). The average grain sizes of ferrite and austenite were found to be 45.7 μm and 36.4 μm, respectively.

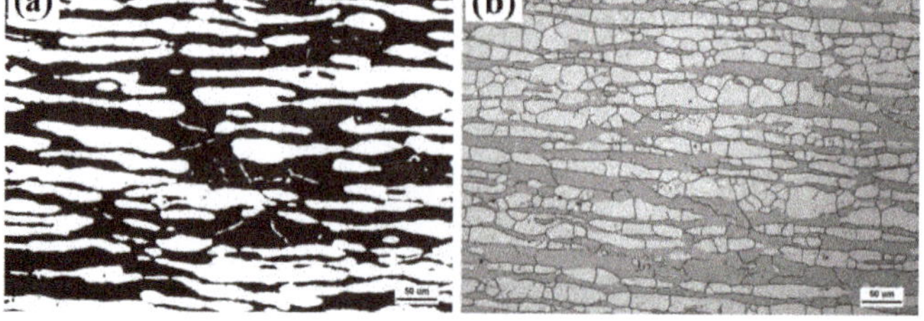


**Figure 2.** Matrix structure of S32750: morphology of the dual-phase structure (**a**); two-phase grain morphology (**b**).

## 3. Results

### 3.1. Surface Morphology Evolution

Figure 3a illustrates the displacement–load curve obtained from the in situ tensile test, with points 1–5 indicating the specific stages at which in situ EBSD measurements were taken.

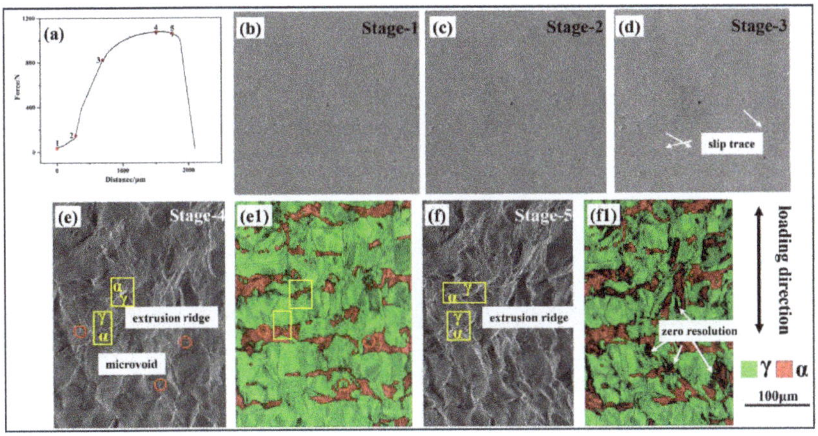


**Figure 3.** Different deformation surface morphologies of S32750 steel: (**a**) displacement-to-force curve; (**b**–**f**) represent stages 1–5; (**e1**) two-phase distribution shown in (**e**); (**f1**) two-phase distribution shown in (**f**).

Figure 3b–f display the surface morphologies corresponding to these five stages. Additionally, (e1) and (f1) depict the phase distribution shown in panels (e) and (f), respectively. It was observed that during the first two stages, the slope of the displacement–load curve is relatively low, indicating that the material is undergoing elastic deformation, with no slip traces visible on the surface. The slope increases markedly in the third stage, signifying that the material has yielded and transitioned into a phase of rapid deformation. Concurrently, distinct glide paths [13] were identified on the surface of the test steel, as referenced in and depicted in Figure 3d. When a crystal is subjected to shear stress, the crystal planes are relatively displaced, leading to permanent deformation and the formation of a slip band [14] on the surface. Such slip bands are commonly observed in face-centered cubic (FCC) structural materials. On the other hand, ferrite contains a large number of screw dislocations, which are more likely to undergo cross-slip [15,16] and move on multiple slip planes. As a result, the formation of slip band structures is relatively rare.

In the fourth stage, as depicted in Figure 3e, the slip bands on the austenite surface tend to become more pronounced and deeper, leading to the formation of micropores and extrusion ridges at the phase boundaries. By the fifth stage, near the point of specimen fracture, there is a notable increase in the size of the extrusion ridges, which bear a resemblance to mountainous terrain, as illustrated in Figure 3f. The emergence and deepening of extrusion ridges and micropores are typically caused by significant plastic deformation and stress concentration within the material, indicating that fracture is imminent. Concurrently, it has been observed that significant deformation occurs at the two-phase boundary, particularly on the side adjacent to the ferrite phase, resulting in the formation of a microscopic black region, as indicated by the arrow in the Figure 3f1. The step size for the electron backscatter diffraction (EBSD) analysis was set at 0.1 μm, which means that microstructures smaller than this dimension could not be resolved with precision, hence appearing as black morphologies in the figure. As a result, duplex stainless steel is highly susceptible to failure at the phase boundaries under sustained tensile stress [17,18], as the continuous deformation can lead to the nucleation and growth of micropores, ultimately causing fracture.

### 3.2. Tensile Fracture Morphology

Figure 4 illustrates the fracture morphology following an in situ tensile fracture. It is evident that the sample's fracture originated from severe plastic deformation on the lower surface, which served as the primary source of cracking.

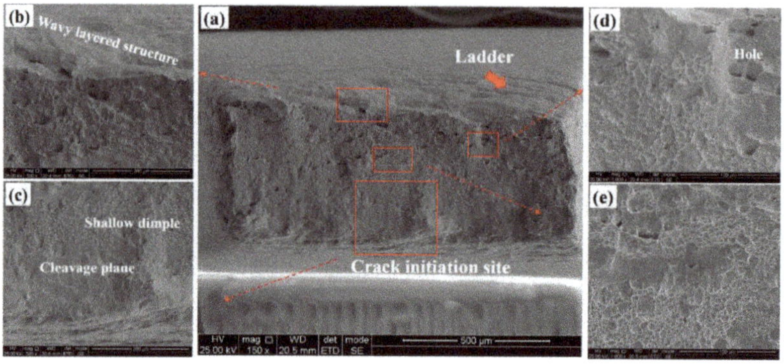


**Figure 4.** Fracture morphology of the in situ tensile specimen: (**a**) macroscopic morphology; (**b**) transient breaking zone; (**c**) crack initiation zone; (**d**) slow propagation zone; (**e**) rapid propagation zone.

This deformation ultimately dictated the final fracture morphology. Additionally, the upper surface exhibits a ladder-like morphology, indicative of the extensive plastic deformation that occurred. Figure 4b presents an enlarged view of the upper surface area, revealing that severe plastic deformation led to the formation of a wavy, layered structure on the specimen's surface. Figure 4c displays an enlarged view of the region proximate to the crack initiation source, characterized by a large cleavage plane and a shallow dimple. The rapid expansion rate during the early stages of crack formation results in the creation of a substantial cleavage plane. As the crack continues to propagate, its rate of expansion decelerates, leading to the progressive enlargement of the dimple morphology, as depicted in Figure 4d,e. Concurrently, the presence of some holes is also observed within this region. This is precisely how the material absorbs energy to slow the rate of crack growth.

The similarities and differences in the fracture morphology between the in situ tensile plate-shaped specimens (Figure 4) and the rod-shaped specimens (Figure 5) were carefully characterized and compared.

Both sample types exhibit surface cracking and the formation of shallow dimples within the crack growth zone. However, the primary distinction lies in the crack propagation paths: plate-like tensile specimens display crack propagation from one end to the other, whereas rod-like tensile samples show cracks initiating from both ends and converging toward the middle. This results in the formation of a larger and deeper dimpled area in the rod-like samples. Additionally, it was observed that the necking in the rod-like samples consistently results in an elliptical shape. This phenomenon can be attributed to the anisotropic mechanical properties of the sample, which become evident after hot rolling. The properties of the sample in the rolling direction and perpendicular to it differ significantly, leading to greater deformation on the sides perpendicular to the rolling direction during the breaking process. Consequently, this results in the formation of an elliptical neck.

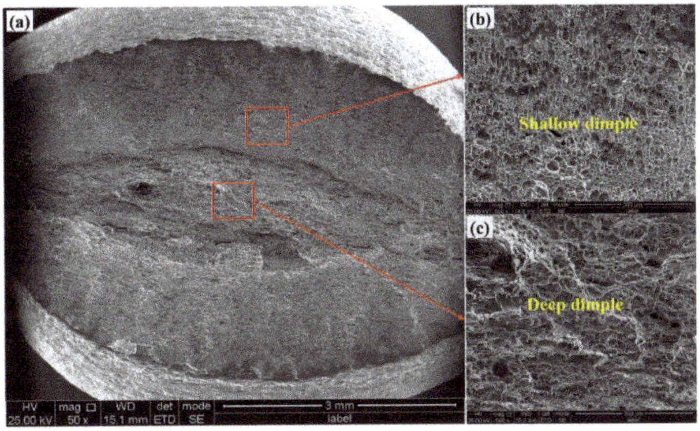


**Figure 5.** Fracture morphology of the rod tensile specimen: (**a**) macroscopic morphology; (**b**) rapid propagation zone; (**c**) slow propagation zone.

### 3.3. Variation in Two-Phase Microhardness

To analyze the strain distribution within the two phases in duplex stainless steel, microhardness measurements were conducted on the ferritic and austenitic phases at various locations. It is understood that stress varies across different positions, and consequently, the strain distribution between the two phases also differs. These measurements provide insight into the strain behavior of both phases across the entire working section. Figure 6a illustrates the relationship between the distance from the fracture and the hardness, while Figure 6b indicates the specific locations where the microhardness tests were conducted. It was observed that the hardness of the two phases does not increase in a strictly monotonic manner; instead, it exhibits a wave-like trend. Notably, the hardness of both phases increases sharply as the distance to the fracture site decreases. Approximately 3.2 mm away from the fracture surface, the microhardness of ferrite is higher than that of austenite.

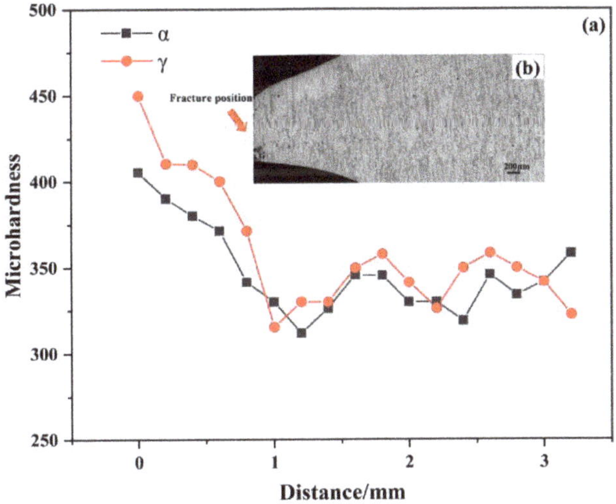


**Figure 6.** Relation between displacement and microhardness: (**a**) microhardness at different distances from the fracture surface; (**b**) the distribution of hardness marks in the matrix structure.

However, as the distance to the fracture site decreases, it is observed that the ferrite phase begins to soften while the austenite phase hardens. This indicates that in the early stages of deformation or when the deformation is minimal, the strain distribution between the two phases is characterized by an increasing strain in austenite and a decreasing strain in ferrite. Within the interval of approximately 1 to 3 mm from the fracture, the hardening behavior of the two phases is nearly identical, exhibiting a wave-like fluctuation. Throughout this region, the microhardness of austenite consistently exceeds that of ferrite. As the distance to the fracture decreases to less than 1 mm, there is a marked increase in the hardness of both phases, with the microhardness difference reaching its peak at the point of final fracture, amounting to approximately 45 units.

## 4. Discussion

### *4.1. Evolution of Two-Phase Microscopic Strain*

The kernel average misorientation (KAM) is a measure that quantifies the average orientation difference in a small unit relative to its neighboring units, primarily reflecting the dislocation activity within a grain. The KAM value is positively correlated with dislocation density, making it a valuable parameter for assessing the extent of plastic strain [19]. KAM values of the two phases can serve as a measure of microscopic damage [20,21]. Higher KAM values typically indicate greater plastic deformation or higher defect density, which are often associated with microscopic damage. Figure 7a through Figure 7e depict the grain boundary distribution in the two phases across the five stages of deformation, while panels (f) through (j) illustrate the corresponding KAM value distributions. It is evident that the KAM values for both phases remain relatively stable during the first three stages. However, upon reaching the fourth stage, there is a notable increase in the KAM values, particularly at the grain and phase boundaries of the two phases. In the fifth stage, near the ferrite side of the two-phase boundary, the strain continues to intensify, leading to the formation of a smaller deformation structure, which appears as the black region in panel (e). This microstructural feature is likely a result of dislocation entanglement at the phase boundary.

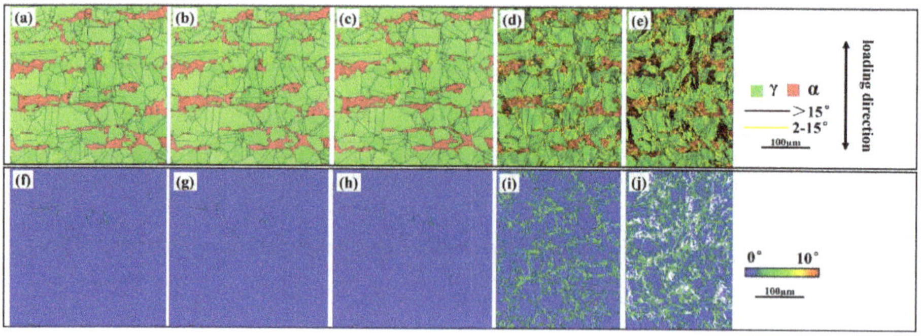


**Figure 7.** EBSD information in the five stages of the S32750 steel deformation (**a**–**e**): 1–5 grain boundary distribution; (**f**–**j**): KAM distribution.

The KAM values for the ferrite and austenite phases were calculated across the five stages of deformation, as shown in Figure 8a. Initially, in the first three stages, the average KAM values remain relatively constant, with the ferrite phase consistently showing a higher KAM value than the austenite phase. However, in the fourth and fifth stages, there is a significant increase in the average KAM values for both phases, suggesting a substantial accumulation of plastic deformation within the ferrite and austenite. By the fifth stage, the KAM value for austenite surpasses that for ferrite, indicating increased dislocation activity within the austenite phase. Concurrently, the accumulation of plastic deformation in ferrite begins to decelerate, suggesting that the ferrite reaches a state of strain saturation earlier than austenite, which retains the capacity for further deformation. As depicted in Figure 8b,c, it is observed that the proportion and growth trend of low-angle grain boundaries (LAGBs) in the two phases, ferrite and austenite, are markedly different.

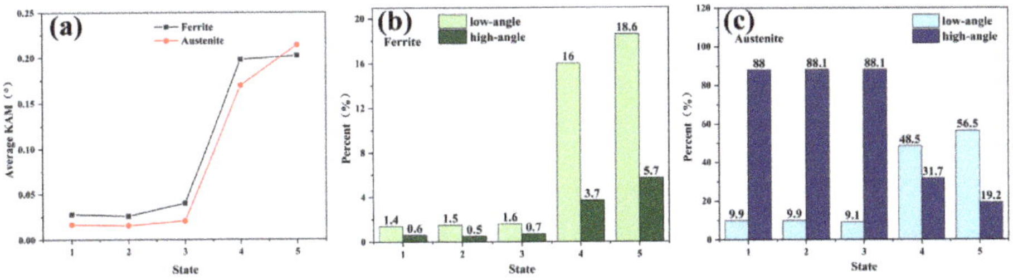


**Figure 8.** Average value of two-phase KAM and grain boundary ratio: (**a**): two-phase KAM averages for the five stages; (**b**,**c**) two-phase grain boundary ratios for the five stages.

During the first three stages of deformation, the number of LAGBs in both phases remains relatively constant. In the ferrite phase, the number of LAGBs exceeds that of high-angle grain boundaries (HAGBs), whereas in the austenite phase, a significant majority—approximately 90%—are HAGBs. In the final two stages of in situ tensile testing, there is a pronounced increase in the number of small-angle grain boundaries within the ferrite phase. In contrast, the austenite phase exhibits a more moderate increase in these grain boundaries. The dislocations in ferrite eventually lead to the formation of small-angle grain boundaries through the cross-entanglement of multiple slip planes, as referenced in [22]. On the other hand, the austenitic phase, characterized by lower frictional shear stress, as noted in [23], tends to accumulate dislocations at the grain and phase boundaries in the form of planar slip, resulting in a less dramatic increase in grain boundary angles.

### 4.2. *Orientation Distribution of the Two Phases*

Figure 9 illustrates the orientation distribution and intensity of the three crystallographic faces for the ferrite and austenite phases in S32750 duplex stainless steel (DSS) in the first and fifth stages of deformation.

Figure 9a,b specifically present the orientation data for the ferrite phase across these two stages. These demonstrate that from the onset to the culmination of deformation, the maximum value of the orientation distribution density for ferrite decreases to approximately half of its initial value, which is around 11. In contrast, as shown in Figure 9c,d, the maximum orientation distribution density of austenite remains constant at both stages, with a value of 4.57. The orientation distribution across the three austenite crystal faces remains relatively uniform, with minimal changes observed even after significant deformation. Conversely, at the conclusion of the deformation process, the grain orientation of ferrite exhibits a clear indication of rotation. In contrast, austenite does not undergo such rotational behavior despite the presence of large plastic deformation. It has also been reported that as stress increases in fatigue experiments, the texture of ferrite changes significantly, while that of austenite remains largely unchanged [24]. Consequently, the ferrite phase encounters increased resistance to deformation along the tensile direction, whereas the austenite phase persists in hardening according to its intrinsic hardening mechanism. This disparity leads to a substantial accumulation of strain on the ferrite side of the phase boundary, ultimately culminating in the specimen's fracture.

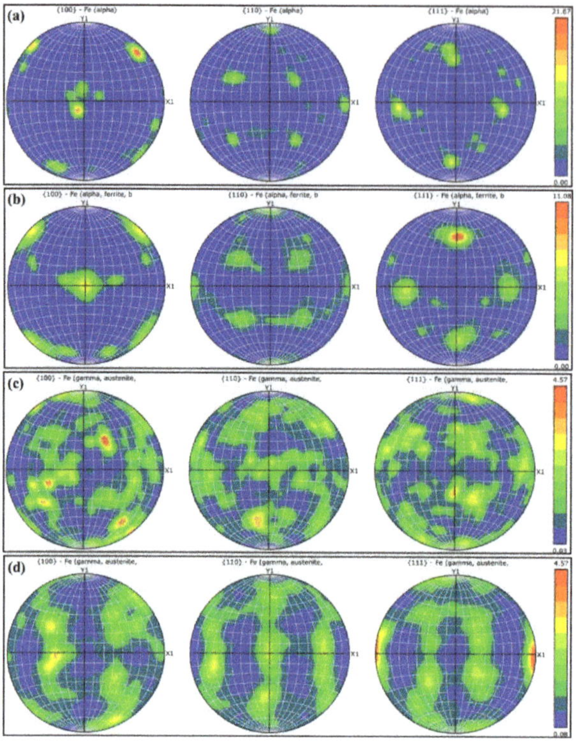


**Figure 9.** Phase orientation distribution in phases 1 and 5: ferrite (**a**,**b**); austenite (**c**,**d**).

## 5. Conclusions

In the present study, the tensile deformation behavior of S32750 duplex stainless steel was investigated using the in situ EBSD technique. Microhardness measurements were employed to assess the hardness variations across the entire region of the specimens. It was observed that the ferrite and austenite phases exhibited distinct deformation mechanisms. The principal conclusions drawn from this research are as follows:

1. Two-phase microhardness results: The microhardness measurements indicate that during deformation, the ferrite phase softens initially, followed by an increase in the hardness of the austenite phase, which eventually surpasses that of ferrite. Near the fracture, within a range of 1–3 mm, the hardness of both phases exhibits a wave-like pattern, suggesting that the deformation coordination between ferrite and austenite occurs through fluctuations in this region. As the distance to the fracture decreases to less than 1 mm, the hardness of both phases increases progressively, with the maximum hardness difference observed near the fracture.
2. The kernel average misorientation (KAM) value for ferrite is higher than that for austenite during the first four stages of deformation; however, this trend reverses in the fifth stage, with the KAM value for austenite surpassing that for ferrite. In the fourth and fifth stages, the small-angle grain boundaries in ferrite grow rapidly, while those in austenite increase at a relatively slower rate. In the later stages of deformation, ferrite undergoes deformation through the rapid formation of a large number of low-angle grain boundaries.
3. The changes in the orientation distribution density across the three crystal faces for the two phases reveal that ferrite grains exhibit a deflection during the deformation

process, whereas austenite grains remain relatively stable. This differential behavior results in continuous straining of the austenite and the accumulation of deformation on the ferrite side of the two-phase boundary. Consequently, this localized deformation leads to the eventual fracture of the material.

**Author Contributions:** S.B. involved in conceptualization, data curation, and writing—original draft preparation; H.F. involved in conceptualization, methodology, and resources; Z.S. involved in supervision and methodology; J.H. involved in investigation and validation; X.W. involved in supervision and writing—review and editing; Y.G. involved in methodology and writing—review and editing. All authors have read and agreed to the published version of the manuscript.

**Funding:** This study was supported by the Hebei Provincial Key Laboratory of Corrosion-Resistant Alloy Materials (Project Number: 23311004D) under the guidance of Han Feng. The project is part of the Science and Technology Program of Hebei Province.

**Data Availability Statement:** The original contributions presented in this study are included in the article. Further inquiries can be directed to the corresponding author.

**Acknowledgments:** I am deeply grateful to Song for his invaluable guidance and constant encouragement throughout this journey of research. His insightful comments and suggestions have played a crucial role in shaping this thesis.

**Conflicts of Interest:** Authors Shun Bao, Han Feng, Zhigang Song, Jianguo He, Xiaohan Wu and Yang Gu were employed by the company Central Iron & Steel Research Institute Co., Ltd. The remaining authors declare that the research was conducted in the absence of any commercial or financial relationships that could be construed as a potential conflict of interest.

## References

1. Li, J.; Shen, W.; Lin, P.; Wang, F.; Yang, Z. Effect of Solution Treatment Temperature on Microstructural Evolution, Precipitation Behavior, and Comprehensive Properties in UNS S32750 Super Duplex Stainless Steel. *Metals* **2020**, *10*, 1481. [CrossRef]
2. Han, Y.; Zhang, Y.; Jing, H.; Gao, Z.; Xu, L.; Zhang, Z.; Zhao, L. Microstructure and Corrosion Studies on Different Zones of Super Duplex Stainless Steel UNS S32750 Weldment. *Front. Mater.* **2020**, *7*, 251. [CrossRef]
3. Gennari, C.; Pezzato, L.; Tarabotti, G.; Zambon, A.; Di Schino, A.; Calliari, I. Influence of Electropulsing Treatments on Mechanical Properties of UNS S32750 Duplex Stainless Steel. *Materials* **2020**, *13*, 1613. [CrossRef] [PubMed]
4. Kangas, P.; Chai, G.C. Use of Advanced Austenitic and Duplex Stainless Steels for Applications in Oil & Gas and Process Industry. *Adv. Mater. Res.* **2013**, *794*, 645–669.
5. Zhang, X.; Yang, C.; Zhu, H.; Xiong, Z.; Cheng, X. Research on deformation behaviors at different temperatures of lean duplex stainless steel S32101 produced by direct cold rolling process. *J. Mater. Res. Technol.* **2024**, *31*, 2900–2914. [CrossRef]
6. Dandekar, T.R.; Kumar, A.; Khatirkar, R.K.; Kumar, D.; Suwas, S. Effect of microstructure and texture on the evolution of mechanical properties in a cold rolled and annealed UNS S32101 lean duplex stainless steel sheet. *Mater. Chem. Phys.* **2023**, *299*, 127485. [CrossRef]
7. Sasaki, S.; Brüggemann, H.; Katsumura, T.; Hirt, G. Hot deformation behavior and microstructure evolution of each single phase in DSS. *J. Mater. Process. Technol.* **2023**, *322*, 118195. [CrossRef]
8. Fu, H.; Dönges, B.; Krupp, U.; Pietsch, U.; Fritzen, C.-P.; Yun, X.; Christ, H.-J. Microcrack initiation mechanism of a duplex stainless steel under very high cycle fatigue loading condition: The significance of load partitioning and micro residual stresses. *Acta Mater.* **2020**, *199*, 278–287. [CrossRef]
9. Zhang, J.; Zhu, Y.; Xi, X.; Xiao, Z. Altered microstructure characteristics and enhanced corrosion resistance of UNS S32750 duplex stainless steel via ultrasonic surface rolling process. *J. Mater. Process. Technol.* **2022**, *309*, 117750. [CrossRef]
10. Angelescu, M.L.; Cojocaru, E.M.; Şerban, N.; Cojocaru, V.D. Evaluation of hot deformation behaviour of UNS S32750 Super Duplex Stainless Steel (SDSS) alloy. *Metals* **2020**, *10*, 673. [CrossRef]
11. Zhao, X.; Liu, C.; Li, Y. Hot working diagram of Super Duplex Stainless Steel UNS S32750. *Hot Work. Technol.* **2015**, *44*, 3.
12. Wu, M.; Hao, W.; Li, J.; Pei, H. Study on dynamic softening Behavior of As-cast Super Duplex Stainless Steel S32750 during Thermal Deformation. *Tisco Sci. Technol.* **2013**, *3*, 32–39.
13. Chai, G.; Peng, R.L.; Johansson, S. Fatigue Behaviors in Duplex Stainless Steel Studied Using In-Situ SEM-EBSD Method. *Procedia Mater. Sci.* **2014**, *3*, 1748–1753. [CrossRef]

14. Zhang, Z.; Li, K.; Cai, T.; Li, P.; Zhang, Z.; Liu, R.; Yang, J.; Zhang, P. Effects of stacking fault energy on the deformation mechanisms and mechanical properties of face-centered cubic metals. *Acta Met. Sin* **2023**, *59*, 467–477.
15. Du, C.; Maresca, F.; Geers, M.; Hoefnagels, J. Ferrite slip system activation investigated by uniaxial micro-tensile tests and simulations. *Acta Mater.* **2018**, *146*, 314–327. [CrossRef]
16. Liang, L.W.; Wang, Y.J.; Chen, Y.; Wang, H.Y.; Dai, L.H. Dislocation nucleation and evolution at the ferrite-cementite interface under cyclic loadings. *Acta Mater.* **2020**, *186*, 267–277. [CrossRef]
17. Rosado-Carrasco, J.; Krupp, U.; López-Morelos, V.; Giertler, A.; García-Rentería, M.; González-Sánchez, J. Effect of a magnetic field applied during fusion welding on the fatigue damage of 2205 duplex stainless steel joints. *Int. J. Fatigue* **2019**, *121*, 243–251. [CrossRef]
18. Alvarez-Armas, I.; Knobbe, H.; Marinelli, M.; Balbi, M.; Hereñú, S.; Krupp, U. Experimental characterization of short fatigue crack kinetics in an austeno-ferritic duplex steel. *Procedia Eng.* **2011**, *10*, 1491–1496. [CrossRef]
19. Choi, J.; Ji, J.; Hwang, S.; Park, K. TRIP aided deformation of a near-Ni-free, Mn–N bearing duplex stainless steel. *Mater. Sci. Eng. A* **2012**, *535*, 32–39. [CrossRef]
20. Nagoshi, T.; Kozu, S.; Inoue, Y.; O'Rourke, B.E.; Harada, Y. Fatigue damage assessment of SUS316L using EBSD and PALS measurements. *Mater. Charact.* **2019**, *154*, 61–66. [CrossRef]
21. Banerjee, A.; Prusty, B.G. Fatigue and fracture behaviour of austenitic-martensitic high carbon steel under high cycle fatigue: An experimental investigation. *Mater. Sci. Eng. A* **2019**, *749*, 79–88. [CrossRef]
22. Wang, W.; Wang, J.; Wang, Q.; Huang, X.; Lu, G.; Liu, Y.; Liu, C. Ferrite-austenite synergistic deformation behavior in a 2205 duplex stainless steel containing equiaxed austenite domains. *Mater. Charact.* **2023**, *205*, 113363. [CrossRef]
23. Dönges, B.; Fritzen, C.P.; Christ, H.-J. Fatigue mechanism and its modeling of an austenitic-ferritic duplex stainless steel under HCF and VHCF loading conditions. In *Fatigue of Materials at Very High Numbers of Loading Cycles;* Springer Fachmedien Wiesbaden: Wiesbaden, Germany, 2018.
24. Zhang, W.; Song, M.; Jiang, W.; Sun, G.; Shan, G.; Chen, M.; Bai, Y. Fatigue failure mechanism of 2205 duplex stainless steel using the neutron diffraction and EBSD technologies. *Int. J. Fatigue* **2022**, *159*, 106828. [CrossRef]

*materials*

MDPI

*Article*

# Material Characterization and Stress-State-Dependent Failure Criteria of AASHTO M180 Guardrail Steel: Experimental and Numerical Investigation

**Qusai A. Alomari [1,2,*], Tewodros Y. Yosef [1,2], Robert W. Bielenberg [1,3], Ronald K. Faller [1,2], Mehrdad Negahban [1,3], Zesheng Zhang [4], Wenlong Li [3] and Brandt M. Humphrey [5]**

1 Midwest Roadside Safety Facility, University of Nebraska-Lincoln, Lincoln, NE 68583, USA; tyosef2@unl.edu (T.Y.Y.); rbielenberg2@unl.edu (R.W.B.); rfaller1@unl.edu (R.K.F.); mnegahban@unl.edu (M.N.)
2 Department of Civil and Environmental Engineering, University of Nebraska-Lincoln, Lincoln, NE 68586, USA
3 Mechanical and Material Engineering, University of Nebraska-Lincoln, Lincoln, NE 68586, USA; wli2@ujs.edu.cn
4 Dimensional Control Systems, Troy, MI 48083, USA; zzszidane@gmail.com
5 Burns and McDonnell Construction, Kansas City, MO 64114, USA; humphrey09@hotmail.com
* Correspondence: qalomari2@unl.edu

**Abstract:** As a key roadside safety feature, longitudinal guardrail steel barriers are purposefully designed to contain and redirect errant vehicles to prevent roadway departure, dissipate impact energy through plastic deformation, and reduce the severity of vehicle crashes. Nevertheless, these systems should be carefully designed and assessed, as localized rupturing, especially near splice or impact locations, can lead to catastrophic failures, compromising vehicle containment, violating crash safety standards, and ultimately jeopardizing the safety of occupants and other road users. Before conducting full-scale crash testing, finite element analysis (FEA) tools are widely employed to evaluate the design efficiency, optimize system configurations, and preemptively identify potential failure modes prior to expensive physical crash testing. To accurately assess system behavior, calibrated material models and precise failure criteria must be utilized in these simulations. Despite the existence of numerous failure criteria and material models, the material characteristics of AASHTO M-180 guardrail steel have not been fully investigated. This paper significantly advances the FE modeling of ductile fracture in guardrail steel, addressing a critical need within the roadside safety community. This study formulates stress-state-dependent failure criteria and proposes advanced material modeling techniques. Extensive experimental testing was conducted on steel specimens having various triaxiality and Lode parameter values to reproduce a wide spectrum of complex, three-dimensional stress-state loading conditions. The test results were then used to identify material properties and construct a failure surface. Subsequent FEA, which incorporated the Generalized Incremental Stress-State-Dependent Damage Model (*GISSMO*) in conjunction with two LS-DYNA material models, illustrates the capability of the developed surface and material input parameters to predict material behavior under various stress states accurately. A parametric study was completed to further validate the proposed models, highlighting their robustness and reliability.

**Keywords:** guardrail steel; ductile fracture; stress-state-dependent failure; triaxiality; lode parameter; GISSMO; LS-DYNA

## 1. Introduction

Longitudinal barriers, such as W-beam steel guardrail systems, are crucial components in roadside safety systems that are widely used throughout the nation's roads and highways to mitigate the severity of run-off-road crashes. Similar to other forms of safety hardware, these guardrail steel systems are essentially designed to contain and redirect errant vehicles while ensuring occupant safety. Specifically, these roadside safety features are purposefully designed to undergo large plastic deformations to properly dissipate the vehicle's impact energy [1]. In the United States, the impact safety performance of guardrail systems is evaluated in accordance with the *American Association of State Highway and Transportation Officials* (AASHTO) *Manual for Assessing Safety Hardware* (MASH) guidelines, with stringent criteria set to assess the crashworthiness and impact behavior of these systems [2,3]. While steel barriers typically exhibit large deflections during vehicle impacts, MASH safety guidelines indicate that vehicles should not penetrate the tested article, implying that guardrails should not sustain major or complete fractures [2]. It has been demonstrated that any localized component failure has the potential to significantly impact the overall performance and crashworthiness of the roadside safety system [4]. Specifically, guardrail steel rupture near the impact or splice locations can result in catastrophic system failure, allowing the errant vehicle to pass through the system [1]. These situations have been highlighted by numerous full-scale crash tests carried out at the Midwest Roadside Safety Facility (MwRSF) at the University of Nebraska-Lincoln [5,6].

For instance, guardrail steel fracture was observed in a full-scale crash test conducted at the MwRSF in 2012 [5]. In the test, a transition from the W-beam Midwest Guardrail System (MGS) to the thrie beam with a concrete curb was evaluated under MASH test number 3–20. Both W and thrie beam sections were manufactured using AASHTO M-180 guardrail steel. In this case, a small passenger car traveling at 101.2 km/h impacted the system at an angle of 25 degrees. As indicated in Figure 1a, the guardrail barrier failed as the vehicle extended and wedged under the W-beam rail, which resulted in severe system damage that encompassed rail rupture at a splice location. Another example of a guardrail fracture occurred recently when the performance of a Bullnose system was evaluated and crash tested [6]. In this test, a small car impacted the thrie beam Bullnose system at a speed of 99.9 km/h. The system failed to satisfy MASH evaluation criteria, as the vehicle was inadequately contained, and windshield deformation exceeded the limits. As illustrated in Figure 1b, this resulted in significant barrier damage, which consisted of extensive bending and deformation, rail tearing, and disengagement of the breakaway posts.

Given the high cost and time duration required to perform full-scale crash testing, Finite Element Analysis (FEA) has emerged as a viable alternative approach for evaluating the crashworthiness and impact performance of roadside safety hardware, such as W-beam guardrail systems [7]. Consequently, FEA is frequently employed to simulate actual collision scenarios to predict barrier performance before conducting expensive full-scale crash testing. Hence, performing high-fidelity FEA is now considered an indispensable stage prior to conducting crash tests to develop optimal system designs. To guarantee reliable computer simulations, an appropriate FE model should be developed and calibrated to accurately mimic the actual behavior of the system [4]. To achieve optimal accuracy and predictive capabilities of the FEA, choosing an adequate material model and corresponding input parameters becomes vital [8]. However, it has been demonstrated that there is no available FE material model fully developed to accurately predict the fracture behavior of AASHTO M-180 guardrail steel [1,9]. One major reason is that the factors contributing to material failure and the associated failure mechanism are not fully understood [10]. Another valid reason is that conducting comprehensive material characterization studies in conjunction with detailed experimental testing to extract material properties under various loading conditions involves considerable effort, substantial cost, and top-tier facilities. The absence of detailed failure models, specifically those that correlate failure to the state of stress, can lead to inaccurate response prediction, which in turn, jeopardizes safety and potentially results in inadequate designs that fail under real-world crash conditions [11–13]. Consequently, alternative models that utilize simplified failure criteria have been frequently used. Nevertheless, these models do not account for damage accumulation or stress-state-dependent failure, which may lead to inconsistent or unreliable material behavior representation.

**(a) MGS to Thrie Beam Transition with Curb System**

**(b) Bullnose with Breakaway Steel Post System**

**Figure 1.** Examples of guardrail steel fractures resulting from full-scale crash testing.

The concept of the stress state is frequently used to represent the interaction between material particles under the influence of complex loading. It was shown by several researchers that the ductile fracture of a material is governed by the state of stress [14–18]. It has long been recognized that failure under a uniaxial state of stress is not typical for metals, as they tend to fail under two- or three-dimensional states of stress [14,17–19]. It has been revealed that ductile fracture of metals is highly dependent on the hydrostatic stress component and the third stress invariant [14–16,20]. More specifically, the magnitude of failure strain is found to be highly influenced by two dimensionless parameters, the stress triaxiality and Lode parameter. While the stress triaxiality defines the degree of hydrostatic stress for any given state of stress, the Lode parameter considers the third stress invariant, leading to a three-dimensional failure representation [17,21,22]. Accordingly, widely accepted failure theories and research efforts adopt this failure criterion to incorporate the effect of complex states of stress in predicting ductile failure of metals [13,23–26].

Considerable research has been devoted to developing an accurate stress-state-dependent ductile failure model. For example, a series of experimental and numerical investigations has been completed by Gao et al. [26] that examines the ductile failure response of DH36 steel plates as a function of stress triaxiality and Lode parameter. Another study carried out by Mohr et al. [4] studied the fracture behavior of Al-7Si-Mg Aluminum alloys. Results from these studies were utilized to develop user-friendly computational material models that can accurately predict fracture behavior. Another study performed by M. Ganjiani investigated the stress-state-dependent fracture performance to propose a FE material model that can be implemented to simulate the response of various metallic alloys under various loading conditions [22]. In the same context, a new ductile failure criterion that accounts for the effects of stress triaxiality and Lode parameter was proposed by Yang et al. [13]. This model was then calibrated against test results of 2024-T351 aluminum alloy, TRIP690 steel, and additively manufactured Ti-6Al-4V. Similarly, numerous research studies have been carried out that examine the fracture of various metals and develop computational constitutive material models [11,19,27–34]. Despite these considerable and extensive efforts, further studies are still needed since the reported findings can only be used to characterize specific materials. Particularly, complete models of ductile failure of many metallic materials, such as AASHTO M-180 guardrail steel, are yet to be constructed. Furthermore, the available material failure models are tailored to encompass loading conditions adequate for sheet forming implementations, which are completely different than those encountered in guardrail steel barriers.

While numerous studies have addressed the ductile fracture behavior of metallic materials using stress-state-dependent criteria [11,13,22,26], to the best of the authors' knowledge, no published or cited study has focused on the complete material characterization and fracture modeling of AASHTO M-180 guardrail steel under vehicular impact conditions. Thus, this study fills a critical gap by providing experimentally calibrated inputs and stress-state-dependent failure criteria tailored to this widely used roadside safety material. Exploring stress-state-dependent ductile failure is crucial to enhance the capabilities of FE simulations and advance the understanding of failure in AASHTO Specification M-180 guardrail steel. As a result, this study aims to develop and calibrate input parameters for advanced LS-DYNA FE material models to accurately simulate the ductile fracture behavior of AASHTO M-180 guardrail steel under complex, vehicle-induced loading conditions. The derived failure surface and material inputs are experimentally validated and implemented to support predictive finite element simulations of roadside safety hardware. The utilized material model is intended to be simple and flexible enough to accommodate any variation in stress state, avoiding any tedious model tweaking and lengthy trial-and-error procedures. The present paper reports the outcomes from several research project stages that involved extensive experimental and numerical efforts [1]. Research outcomes are expected to be invaluable resources for the roadside safety community as they will enhance the current understanding of guardrail performance and significantly contribute to its practical applications.

## 2. Preliminary Knowledge

The concept of the stress state is of paramount significance in the fields of mechanics and material science as it focuses on the internal forces that particles in a material exert on each other under the influence of external loads. Hence, understanding the state of stress is pivotal in predicting the behavior of a material under various combinations of loading scenarios, and, more importantly, facilitating the prediction of potential failure. In general, a tensor is a mathematical representation of a physical property. In continuum mechanics, the stress tensor ($\sigma$) is a fundamental concept that is used to define the state of stress at any point in a body [35]. As expressed in Equation (1), a true stress tensor comprises two distinct components, hydrostatic stress ($\sigma_H$) and deviatoric stress ($s$) [24]. Here, the hydrostatic stress tensor is defined by the product of the trace of a tensor (i.e., $tr(\sigma)$ the sum of its diagonal components), and $\boldsymbol{I}$, which is the second-order identity tensor. The hydrostatic stress component represents the isotropic portion of the stress tensor, which corresponds to the uniform pressure experienced by a material [36,37]. Hydrostatic stress is also equivalent to the negative value of the uniform pressure ($p$) being exerted on the particle of the material. On the other hand, deviatoric stress represents the shear stresses.

$$\sigma = s + \sigma_H \boldsymbol{I} = s + \frac{1}{3} tr(\sigma)\boldsymbol{I} = s - p\boldsymbol{I} \tag{1}$$

It is well known that ductile materials undergo large plastic deformations before failure [13]. These deformations can be attributed to the volumetric and shape changes. In general, hydrostatic stress induces volumetric changes, while deviatoric stress triggers shape deformations [36,37]. For isotropic materials, stress tensors can also be expressed as three stress invariants ($I_1$, $I_2$, $I_3$) [13]. As shown in Equations (2)–(4), these invariants are utilized to provide other means to describe the state of stress [20,21]. Another basic concept in ductile fracture and mechanics of materials is the von Mises or equivalent stress ($\sigma_m$), as represented in Equation (5). In this equation, the operator $s : s$ represents the double contraction of the deviatoric stress tensor with itself, which corresponds to the second invariant of deviatoric stress (i.e., $s : s = \sum_{i,j} s_{ij}^2$). Typically, the von Mises stress corresponds to the equivalent shear stress and is commonly used to identify the yielding of a material [38]. It has been indicated that void growth and coalescence, which control the microscopic ductile failure of metals, are mainly influenced by the Von Mises stress ($\sigma_m$) [13,38]. The Von Mises stress can also be expressed in terms of the second invariant ($J_2$) of the deviatoric stress, which refers to distortional stress resulting from the yielding of an isotropic ductile material.

$$I_1 = tr(\sigma) \tag{2}$$

$$I_2 = \frac{1}{2}\left[(tr(\sigma))^2 + tr\left(\sigma^2\right)\right] \tag{3}$$

$$I_3 - det(\sigma) \tag{4}$$

$$\sigma_m = \sqrt{\frac{3}{2} s : s} = \sqrt{3J_2} \tag{5}$$

Numerous research efforts have demonstrated that the ductile fracture of metals is highly dependent on the hydrostatic stress component [14–16,20]. Initially, Bridgman reported that the ratio between the initial cross-sectional area and the area after fracture at the neck of a round steel bar being tested under uniaxial tension is directly related to the induced confining pressure [14]. In fracture mechanics, stress triaxiality ($\eta$) is a dimensionless parameter that is frequently used to identify the degree of hydrostatic stress for any given state of stress [21]. As demonstrated in Equation (6), stress triaxiality is defined as the ratio between hydrostatic stress and von Mises stress, which is an indirect representation of the first and second stress invariants [17,26,35]. Currently, stress triaxiality is widely employed among researchers to investigate the fracture behavior of ductile metals [13,17,20,22].

Although stress triaxiality has been widely recognized as a key factor in the development of ductile damage, numerous research studies have shown that it is not the only determinant, particularly when the material experiences complex loading conditions [13,16,17,22,25,26]. Rather than only using two of the three invariants of the state of stress, information from all three invariants is needed to describe the state of stress in crash and ballistic impact applications. To extend the failure model to capture all the characteristics of the state of stress for an isotropic material, it was indicated that incorporating the third stress invariant as an additional parameter is essential to establish a full characterization of triaxiality [16,17]. It was revealed in these studies that the Lode parameter ($\xi$), which is commonly referred to as the normalized third deviatoric stress invariant ($J_3$), influences the fracture of ductile metals in addition to stress triaxiality and should be taken into account. Lode parameter and the third deviatoric stress invariant are given in Equations (7) and (8). Following relevant previous research studies and given that this study intends to examine the ductile fracture of AASHTO M-180 guardrail steel under impact loading, stress triaxiality and the Lode parameter were employed to establish the stress-state-dependent ductile failure criteria.

$$\eta = \frac{\sigma_H}{\sigma_m} = -\frac{p}{\sigma_m} \tag{6}$$

$$\xi = \frac{27}{2}\frac{J_3}{\sigma_m} \tag{7}$$

$$J_3 = det(s) \tag{8}$$

## 3. Experimental Testing and Material Characterization

### *3.1. Material Selection*

Given the challenges associated with extracting round and complex shapes of test specimens from standard W-beam rail sections, identifying a substitute material that precisely represents the mechanical and chemical properties of AASHTO M-180 guardrail steel and can be readily processed and formed is pivotal. The AASHTO M-180 steel specification outlines the minimum acceptable mechanical properties, such as yield strength, ultimate tensile strength, and elongation, as detailed in Table 1. A survey of guardrails available in the market was conducted to establish a range of mechanical properties deemed suitable, with material certificates and properties compiled in a recent study by Schmidt et al., facilitating the computation of median, 15th, and 85th percentile values for yield strength, tensile strength, and elongation [39]. This data was instrumental in setting reasonable bounds during the material search. Furthermore, the AASHTO standards do not specify a particular ASTM specification that the guardrail should comply with, other than requirements concerning bolts, nuts, and Zinc coatings. This underscores the critical role of selecting a steel specification that delineates the material required for the component testing program. In accordance with the findings of Schrum et al. [40], it was concluded that ASTM A572 steel can closely resemble AASHTO M-180 guardrail steel available in the market in terms of chemical and mechanical properties, making it a feasible substitute material option for the component testing program [41]. ASTM A572 Grade 50 steel plate of the designated material with 0.5 inches (12.7 mm) thickness and 60 inches (1524 mm) width and with properties comparable to the median values of the range of materials accessible in the market was used, as demonstrated in Table 1. It is worth noting that this selected material falls within the 15th and 85th percentile ranges for both yield and ultimate strength, while demonstrating an elongation that slightly exceeds the 85th percentile, which was deemed acceptable as it falls within the range of compiled information on material available in the market.

**Table 1.** AASHTO M-180 and selected material mechanical properties.

| Material | Yield Strength ksi (MPa) | Ultimate Strength ksi (MPa) | Elongation (% in 2 in.) |
|---|---|---|---|
| AASHTO Specification M-180 Minimum Values | 50 (345) | 70 (483) | 12% |
| Median Values of Compiled Material Available on the Market | 63.2 (436) | 74.8 (516) | 27.1% |
| Selected Material Mean Values from Material Certificate | 65 (448) | 74 (510) | 30.5% |

### 3.2. *Development of Component Testing Program*

It has been demonstrated that failure in guardrail steel under vehicle impact predominantly occurs under complex three-dimensional states of stress, characterized by a wide range of stress triaxiality and Lode parameter values that influence ductile fracture behavior [19,26,30]. Typically, the effects of stress state on ductile fracture can be expressed through specially designed experiments that involve standard and non-standard test specimens [26]. Accordingly, the configuration and loading pattern of test specimens must be carefully selected to represent distinct stress triaxiality, Lode parameter, and effective plastic strain at failure [11]. Building upon these insights, a material testing program was developed assuming that the selected material, representing AASHTO M-180 guardrail steel, will fail at an effective plastic strain depending on the state of stress. Following several research studies, 21 test specimens with different geometries were selected and tested under tension, compression, shear, punching, and torsional loading conditions [16,17,20,25,26,42,43].

#### 3.2.1. Tension Specimens

Thirteen specimen geometries (specimens no. 1 through no. 13), depicted in Figure 2, were tested under tension. The first geometry refers to a flat dog bone specimen, which is a standard specimen geometry that is consistently used to perform uniaxial tension tests. This specimen represents the reference stress state for all other specimens and can be used to extract basic material properties, including elastic modulus and yield strength [1]. The specimen is characterized by $\eta = 0.333$ and $\xi = 1$. Specimens no. 2 through no. 4 correspond to flat notched geometries that are commonly used to examine the effects of varying stress states associated with plane stress conditions. In this set of specimens, it was indicated that hydrostatic forces developed in the neck region vary with the notch diameter [14]. Particularly, lateral confinement stress increases with reducing the notch diameter, resulting in various stress states along the reduced cross section of the specimen. Thus, these specimens are attributed to $\eta = 0.40$ to 0.577 and $\xi = 0$ to 1. Similar to specimen no. 1, an axially symmetric, round, smooth specimen is another alternative geometry to perform simple uniaxial tension tests and corresponds to the same set of triaxiality and Lode parameter. Specimens no. 6 through no. 10, identified by notched round geometries, were included in the test program. Again, varying the radius of the notch produces various stress states characterized by $\eta = 0.40$ to 0.80 and $\xi = 1$. Specimens no. 11 through no. 13 are featured by plane strain conditions, with $\eta = 0.577$ to 0.80 and $\xi = 0$. In these specimens, thickness and notch size are designed so that the strain in the direction perpendicular to the loading direction is negligible, simulating a plane strain condition. It was indicated that specimens with a thickness that exceeds 25.4 mm can accurately mimic theoretical plane strain conditions when tested under tension [17,25].

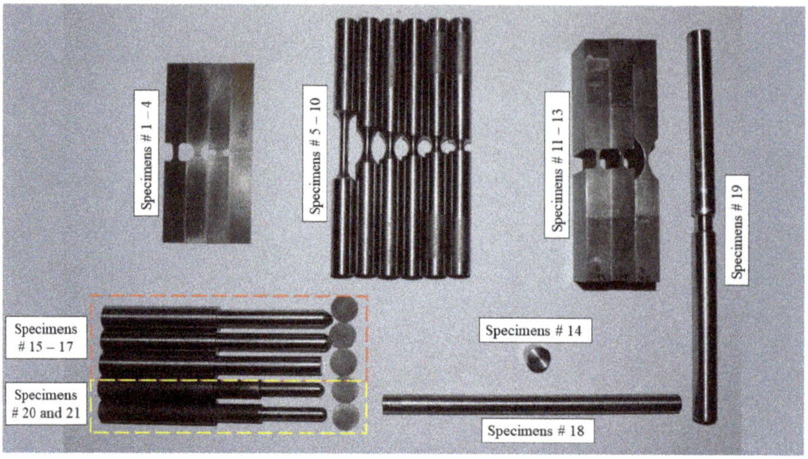


**Figure 2.** Test specimens.

#### 3.2.2. Compression and Punch Specimens

Compression and punching test specimens, depicted in Figure 2, were also considered in the test program to provide a diverse range of possible loading scenarios. Specimen no. 14, which refers to the cylindrical upsetting geometry, was tested under uniaxial compression, representing $\eta = -0.333$ and $\xi = -1$. Five punch head shape specimens, associated with specimens nos. 15 through 21, were manufactured and used to test circular thin steel specimens. Punch head shapes and specimen geometries matched those investigated by Buyuk and Seidt and further revised by Humphrey [1,25,43]. Specimens with round and standard heads correspond to $\eta = 0.5774$ to 0.667 and $\xi = -1$, while sharp heads are attributed to $\eta = 0$ and $\xi = 0$.

#### 3.2.3. Shear and Torsion Specimens

Shear and torsion specimens were fabricated and considered in the test program. A Dual-Point Shear Fixture MTS system was employed to test a solid steel rod specimen (specimen no. 18) under shear loading, representing a zero triaxiality and Lode parameter state of stress. A thin hollow tube sample (specimen no. 19) was fabricated and tested under torsion. Both specimen configurations are essentially designed to represent the state of zero triaxiality and Lode parameter.

### *3.3. Testing Setup and Instrumentation*

Tension, compression, shear, and punching tests were completed using two material testing systems (MTS) servo-hydraulic universal testing machines, illustrated in Figure 3. Landmark load frames include a 22-kip (100 kN) fatigue testing frame and a 220-kip (1000 kN) criterion static testing frame. On the other hand, torsion tests were carried out using a Tinius Olsen & Company (Horsham, PA, USA) of Philadelphia torsion frame. Displacements and strains for each specimen were measured using a 1- and 2 in.-gauge length MTS axial extensometer (MTS Systems Corporation, Eden Prairie, MN, USA), the LX 500 laser extensometer (MTS Systems Corporation, Eden Prairie, MN, USA). Following published material testing reports and relevant studies, three-dimensional digital image correlation (3D-DIC) was also employed in the component testing program, as full strain-displacement fields can be accurately determined throughout the specimen loading stages, and more importantly, beyond the necking point [1,25,43]. Additional details on the testing setup and instrumentation can be found elsewhere [1].

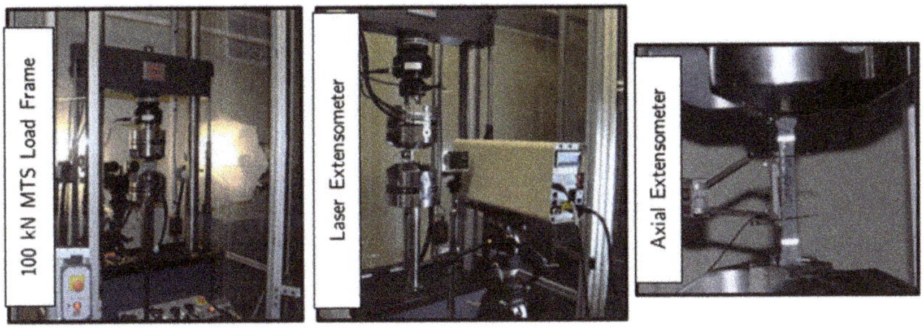


**Figure 3.** Overview of the tensile test instrumentations and setup.

### *3.4. Testing Matrix*

A total of 124 tests were carried out on the 21 specimens illustrated in Figure 2. While some test results were utilized to extract material properties, the remaining tests were used to provide a diverse range of stress states for evaluating the failure criterion. For instance, the DIC technique was employed to test one standard flat dog bone specimen (specimen no. 1) to identify the elastic modulus and Poisson's ratio of the tested surrogate material. Five additional tests on specimens no. 1 and no. 5 were conducted using laser and axial extensometers to determine yield strength and the isothermal hardening curve. The complete plot of the targeted stress states, as represented by triaxiality and Lode parameter, for each specimen type is illustrated in Figure 4.

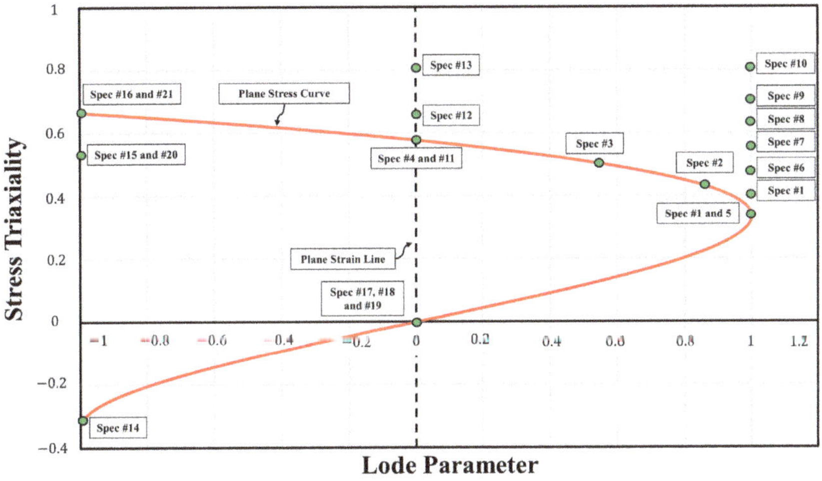


**Figure 4.** Target stress states of testing specimens.

### *3.5. Testing Results*

Representative engineering stress–strain curves for specimens no. 1 through no. 14 are illustrated in Figures 5 and 6. It should be noted that at least two distinct tests were performed on each specimen configuration. In each of these tests, slight variations in stress–strain plots were observed. The variation can be attributed to several factors, such as displacement and force measurement methods, and other test uncertainties. Consequently, the plots depicted in these figures correspond to the curves lying between the upper and lower bounding curves to provide a more realistic representation of stress–strain relationships.

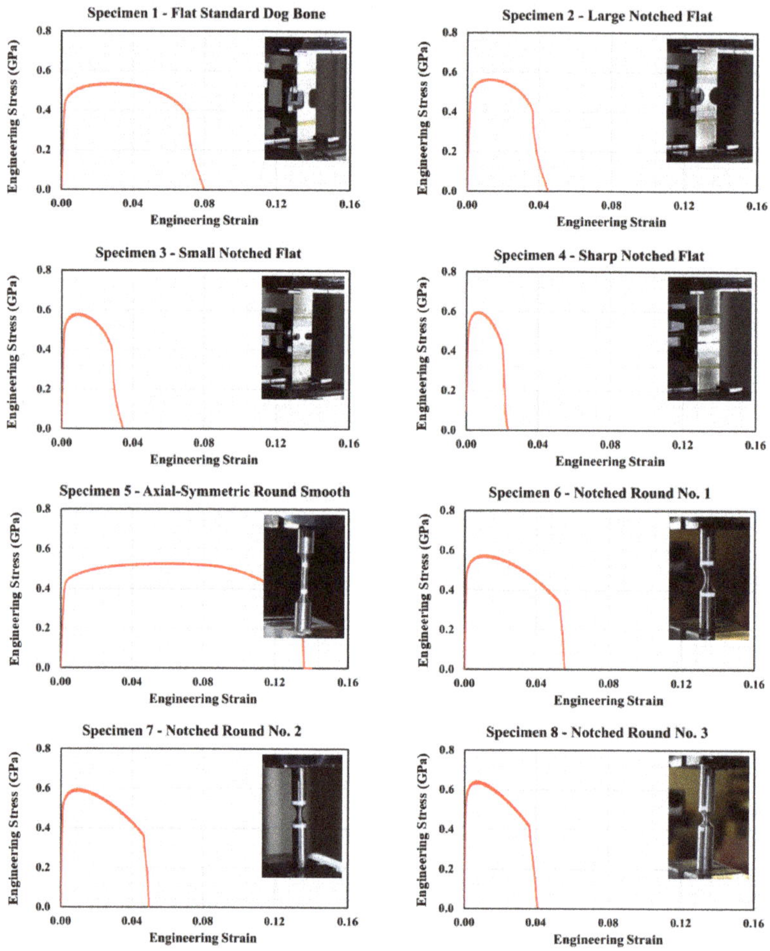


**Figure 5.** Engineering stress–strain plots for specimens 1–8.

As shown in Figure 5, specimen no. 1 demonstrated typical ductile behavior characterized by an initial elastic region followed by substantial plastic deformation. As expected and due to stress concentration at the notch, specimens no. 2 through no. 4 demonstrated lower failure strain values compared to specimen no. 1. In particular, reducing the notch size contributed to increasing stress triaxiality, which resulted in promoting localized deformation and decreasing the overall ductility. Similar to specimen no. 1, the axially symmetric, round, smooth specimen (specimen no. 5) experienced consistent stress distribution with high ductility. Again, the influence of altering notch size and the corresponding higher levels of triaxiality is further highlighted in the stress–strain response of specimens no. 6 through no. 10, where reduced d levels of ductility were observed. Specimens no. 11 through no. 13, which induce a complex state of stress simulating plane strain loading conditions, demonstrated higher resistance to necking due to their relatively larger cross-sectional areas, and thus, high ultimate strengths. While specimen 11 sustained a relatively extended plastic deformation range before failure, the two other specimens demonstrated reduced ductility similar to the flat and round notched specimens. When tested under compression, specimen no. 14 has shown a different failure mechanism, which is characterized by barreling and ultimately shear failure. As shown in Figure 6, the resulting stress significantly increased prior to failure due to the lateral confinement effects. Results from punch and shear tests are illustrated in Figure 7. This figure clearly illustrates the effect of punch head shape on specimen capacity and failure strain. Due to unforeseen technical errors during testing, the force-displacement plot corresponding to specimen no. 19 is not reported herein. A comprehensive summary of component testing results is provided in Tables 2 and 3.

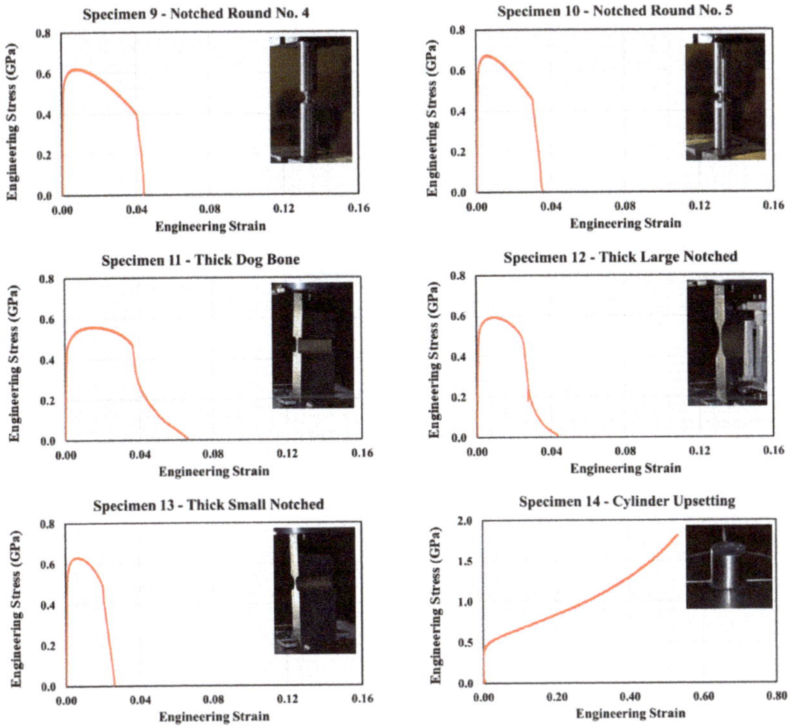


**Figure 6.** Engineering stress–strain plots for specimens 9–14.

**Table 2.** Summary of test results, specimens 1–13.

| Specimen No. | Yield Strength (MPa) | Ultimate Strength (MPa) | Effective Strain at Failure (DIC) |
|---|---|---|---|
| 1 | 444 | 547 | 0.82 |
| 2 | 488 | 567 | 0.63 |
| 3 | 499 | 582 | 0.76 |
| 4 | 533 | 615 | 0.54 |
| 5 | 432 | 527 | 1.45 |
| 6 | 489 | 567 | 1.22 |
| 7 | 498 | 595 | 1.13 |
| 8 | 530 | 624 | 1.23 |
| 9 | 546 | 640 | 0.82 |
| 10 | 587 | 678 | 1.40 |
| 11 | 445 | 561 | 1.04 |
| 12 | 503 | 599 | 0.98 |
| 13 | 538 | 633 | 0.57 |

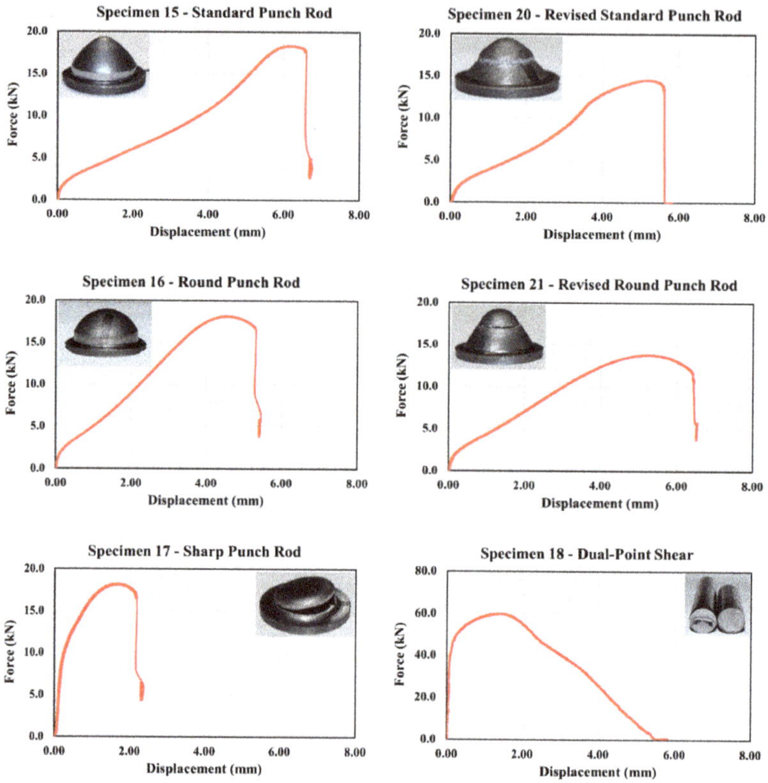


**Figure 7.** Force-displacement plots for specimens 15–18, 20, and 21.

**Table 3.** Summary of test results, specimens 14–18, 20, and 21.

| Specimen No. | Maximum Force (kN) | Maximum Displacement (mm) |
|---|---|---|
| 14 | 250 | 7.165 |
| 15 | 18.7 | 6.330 |
| 16 | 18.2 | 4.536 |
| 17 | 18.2 | 1.666 |
| 18 | 59.6 | NA |
| 20 | 13.3 | 5.552 |
| 21 | 13.8 | 5.248 |

### *3.6. Elastic and Plastic Flow Parameters Extraction*

Accurately extracting elastic and plastic material properties from experimental testing and adopting an appropriate elastoplastic material model are the primary steps to guarantee reliable simulations of guardrail steel behavior, specifically the fracture mechanism [19]. Uniaxial tensile tests carried out on standard tensile specimens, which are fabricated per universal specifications, such as ASTM E8, are typically used to identify the mechanical properties of metals [41,44]. Tension tests are commonly used to identify major material parameters, including yield strength ($\sigma_y$), ultimate tensile strength ($\sigma_u$), elastic modulus ($E$), Poisson's ratio ($\nu$), and failure strain ($\varepsilon_f$). These parameters are evaluated through establishing a constitutive relationship between the applied tensile load and the specimen's deformation, which is commonly expressed in the form of force-displacement or stress–strain curves. In this study, tests carried out on specimen no. 1, which represented uniaxial tension, were used to identify the elastic material properties. The three-dimensional digital image correlation (3D-DIC) technique was employed to accurately measure variations in the axial and transverse strains. Accordingly, the elastic modulus and Poisson's ratio were found to be 199.3 GPa and 0.295, respectively. While these properties are essential for simulating the material's initial elastic response and response during unloading, post-yielding, plastic behavior requires characterization of the flow parameters. Engineering and true stress–strain relationships are widely accepted among researchers to represent the behavior of metals [35]. Engineering stress–strain curves correspond to the initial undeformed configuration when calculating stress from load. In uniaxial extension, engineering stress ($\sigma$) and engineering strain ($\varepsilon$) are expressed by Equations (9) and (10), where $F$ is the applied tensile force, $A_o$ is the initial cross-sectional area, and $l_o$ and $l_f$ are the initial and final gauge lengths, respectively.

$$\sigma = \frac{F}{A_o} \tag{9}$$

$$\varepsilon = \frac{l_f - l_o}{l_o} \tag{10}$$

While engineering stress–strain curves is useful for initial material characterization, they become less accurate when a material undergoes large deformations. This is because they do not account for the reduction in cross-sectional areas and the actual local elongation of the specimen, as they assume constant cross-sectional area and uniform elongation throughout the test. Accordingly, utilizing alternative stress–strain measurements is pivotal to provide a more accurate representation of local material response. Hence, the true stress–strain relationship is frequently utilized as it considers an instantaneous cross-sectional area. True stress–strain curves are subsequently adopted in numerous implementations that involve large plastic deformations, such as metal forming and roadside safety applications. True stress and true strain are given in Equations (11) and (12). It is noteworthy to mention that engineering and true stress–strain curves are close to the yielding point, and the discrepancy significantly increases as the material experiences large plastic deformations. Once the diffused necking begins, these equations are no longer applicable. This is mainly because the uniaxial state of stress is transformed to a triaxial stress state, leading to nonuniform plastic deformation [44,45].

$$\sigma_T = \sigma\,(1 + \varepsilon) \tag{11}$$

$$\varepsilon_T = \ln(1 + \varepsilon) \tag{12}$$

Failure strain can be calculated using the adjusted method proposed by Bridgman [14], as illustrated in Equation (13), where $A_i$ and $A_f$ are the initial and final cross-sectional areas, respectively. On the other hand, diffused necking occurs when the true stress is equal to its hardening rate, which refers to Considère's necking criterion, as shown in Equation (14) [19]. Graphically, the necking stress is located at the intersection between the true stress–strain curve and its derivative.

$$\varepsilon_F = \ln\left(\frac{A_i}{A_f}\right) \tag{13}$$

$$\sigma_T = \frac{d\sigma_T}{d\varepsilon_T} \tag{14}$$

In FE simulations, the input of true stress–strain data that precisely represents the plastic flow of a material is essential to ensure the accuracy of the results. Although linear interpolation is frequently used to define post-necking true stress–strain behavior, the process is quite tedious, as it involves several trials to achieve an adequate match between simulation and test results. As the ultimate goal of the current study is to provide consistent material characteristics that can be consistently adopted to model AASHTO M-180 guard rail steel, an alternative and more feasible approach to characterize the post necking behavior was implemented. Moreover, the selected strain hardening model should be robust and stable when various specimen configurations are numerically investigated. Numerous mathematical models to estimate post-necking true stress–strain are available in the open literature [46]. Following a preliminary FE simulation stage, the Swift model, which is extensively used to model isotropic hardening of metals, was adopted in the present study to represent the material's plastic flow [47]. The model can be used to express the true stress as a function of true (or effective) plastic strain ($\varepsilon_P$), which can be obtained by subtracting the elastic strain portion from the total strain value, as expressed in Equation (15). The Swift model is provided in Equation (16), where $K$, $\varepsilon_o$, and $n$ are material constants.

$$\varepsilon_p = \varepsilon_T - \frac{\sigma_T}{E} \tag{15}$$

$$\sigma_T = K\left(\varepsilon_o + \varepsilon_p\right)^n \tag{16}$$

Initially, the true stress–strain response associated with specimen no. 5 was extracted. 3D-DIC was employed to estimate the true axial strain and the subsequent effective plastic strain. Non-linear regression analysis was implemented, using IBM SPSS software version 25, to fit the Swift model with the measured true stress–strain data up to the necking point. Following multiple iterations, optimized Swift model parameters were found to be $K$ = 0.8 GPa, $\varepsilon_o$ = 0.01, and $n$ = 0.13. The final Swift model was then used to define the full relationship between true stress and effective plastic strain (i.e., beyond necking and until failure). The final proposed plasticity curve is shown in Figure 8.

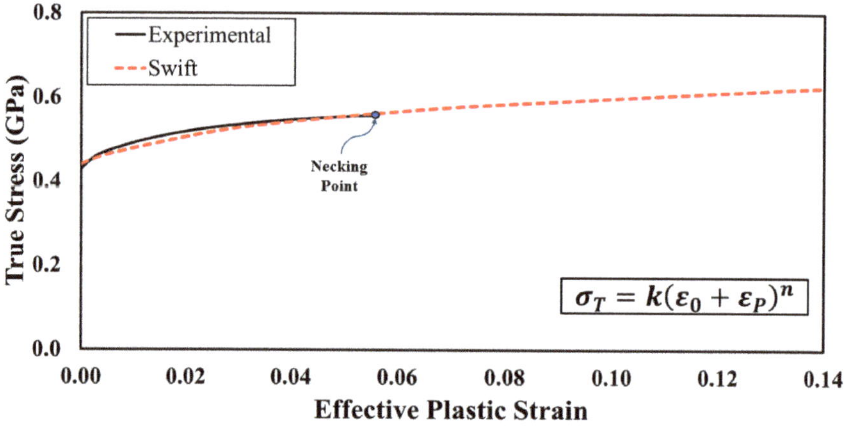


**Figure 8.** Derived plasticity curve.

## 4. Finite Element Analysis

### *4.1. Material Models*

This section presents a brief overview of the three selected LS-DYNA 14.0 material models commonly used to model ductile failure. These models were selected due to popularity and proven effectiveness in simulating material responses in crashworthiness applications.

#### 4.1.1. Tabulated Johnson–Cook Model

An extensive collaborative effort involving the Federal Aviation Administration (FAA), National Aeronautics and Space Administration (NASA), George Washington University (GWU), Ohio State University (OSU), and George Mason University (GMU) has led to the development of the Tabulated Johnson–Cook (*MAT224*) constitutive material model in LS-DYNA 14.0 for simulating metals [48]. This model has been recognized by several researchers as a viable tool for various practical applications, including aerospace and roadside safety applications [1,48]. The Tabulated Johnson–Cook model builds on the existing Johnson–Cook material model (*MAT15*) in LS-DYNA 14.0, with the added capability of incorporating tabulated input parameters. Beyond defining the material's density and elastic properties, this elastic thermo-viscoplastic material model allows for the input of multiple stress–strain curves under varying strain rates and temperatures, as expressed in Equation (17) [49]. In this equation, $\sigma$ is the flow stress, $A$ is the yield stress at a reference temperature and strain rate, $B$ is the coefficient of strain hardening, $n$ is the strain hardening exponent, $\varepsilon^p$ is the effective plastic strain, C and $m$ are material constants, $T^*$ is the homologous temperature, which is given by Equation (18) and relates the material's temperature to the room and melting temperatures.

$$\sigma = \left(A + B\varepsilon^{pn}\right)\left(1 + \mathrm{C}\ln\dot{\varepsilon}^{P}\right)\left(1 - T^{*m}\right) \tag{17}$$

$$T^* = \frac{T - T_{room}}{T_{melt} - T_{room}} \tag{18}$$

In *MAT224*, the effective plastic failure strain ($\varepsilon^{P}{}_{f}$) is defined as a function of several critical parameters: stress triaxiality ($\eta$), Lode parameter ($\xi$), total or plastic strain rates ($\dot{\varepsilon}$ or $\dot{\varepsilon}^{P}$), temperature ($T$), and element size ($l_o$) [1]. It is important to note that failure strain can be specified as a function of one or more of these parameters. When multiple parameters are employed, the failure strain is calculated as the product of these functions, as shown in Equation (19). To accurately model stress-state-dependent failure, the latest version of LS-DYNA 14.0 enables the input of a curve that defines the effective failure strain as a function of stress triaxiality. Additionally, for solid elements, a table of curves, which is a failure surface, can be utilized to relate the failure strain to both stress triaxiality and Lode parameter. The load curve or table ID can be specified using the *LCF* variable on the second input card. Furthermore, the *LCG* and *LCH* variables are used to assign load curve IDs, where each curve defines failure strain as a function of strain rate or temperature. A specific curve or table ID can be entered under *LCI* to correlate the failure strain to element size and stress triaxiality. It is noteworthy that the stress triaxiality referenced in all input curves corresponds to the ratio of pressure (i.e., $-\sigma_H$) to von Mises stress.

$$\varepsilon^{P}{}_{f} = f(\eta,\, \xi)g\left(\dot{\varepsilon}^{P}\right)h(T)i(\eta, l_o) \tag{19}$$

The Tabulated Johnson–Cook material model incorporates two built-in failure criteria. The default failure criterion is based on the accumulation over time of the effective plastic strain rate over the effective failure strain, as shown in Equation (20). This approach accounts for load-path dependent failure. For modeling load-path independent failure, an alternative criterion can be activated (*FAILOPT* = 1.0), as outlined in Equation (21), where only the current state of plastic strain is evaluated. When this criterion is enabled, the effective failure plastic strain is averaged over several time steps, which can be controlled using the *NUMAVG* variable. Element deletion occurs when either $F$ or $F_2$ is reached or exceeded unity.

$$F = \int \frac{\dot{\varepsilon}^P}{\varepsilon^P{}_f} dt \tag{20}$$

$$F_2 = \frac{\varepsilon^P}{\varepsilon^P{}_f} \tag{21}$$

4.1.2. Generalized Incremental Stress-State-Dependent Model (GISSMO)

Several LS-DYNA 14.0 material models have been developed to address material damage and failure [50]. Many of these models combine elastoplasticity with damage formulations and embedded failure criteria, such as the Johnson–Cook models (*MAT15* and *MAT224*). On the other hand, LS-DYNA 14.0 supports the definition of damage and failure through the **MAT_ADD_EROSION* keyword [50]. This keyword allows users to specify a failure model, enabling the integration of distinct failure criteria with the chosen material model. One frequently used model for predicting failure based on damage accumulation is the Generalized Incremental Stress-State-Dependent Model (*GISSMO*), which can be implemented via the *MAT_ADD_EROSION* keyword [23]. Originally developed by Neukamm and Haufe in 2010 and further refined by Andrade in 2016 for crashworthiness applications, *GISSMO* provides a robust framework for simulating material failure [24,51,52]. Notably, LS-DYNA also offers the *MAT_ADD_DAMAGE_GISSMO* keyword, a variant of the *MAT_ADD_EROSION* keyword, which facilitates the direct implementation of the *GISSMO* failure model. When *GISSMO* is incorporated, damage accumulation ($\Delta D$) is calculated at each time step according to Equation (22). In this equation, *DMGEXP* refers to the damage exponent, $\varepsilon_f(\eta,\xi)$ is the stress state dependent failure strain (i.e., function of the stress triaxiality and Lode parameter), $D$ is the current damage value, and $\Delta\varepsilon_p$ is the equivalent plastic strain increment [49]. Similar to *MAT224*, a load curve ID that defines failure strain as a function of stress triaxiality, or a table ID that accounts for the effects of stress triaxiality and the Lode parameter, can be specified using the *LCSDG* variable [49]. Unlike *MAT224*, stress triaxiality is defined according to Equation (6), where hydrostatic stress is divided by the von Mises stress. Element deletion is triggered when the damage value of $D$ reaches or exceeds unity.

$$\Delta D = \frac{DMGEXP}{\varepsilon_f(\eta,\xi)} D^{\left(\frac{DMGEXP-1}{DMGEXP}\right)} \Delta\varepsilon_p \tag{22}$$

Through the *GISSMO* model, an additional parameter, $F$, which serves as an instability measure defining the onset of diffused necking, can be activated [53]. When this instability variable is used, coupling between damage and stress flow will be initiated and evaluated at each time step, as shown in Equation (23) [24]. In this context, $\varepsilon_{crit}(\eta,\xi)$ represents the critical strain at the onset of diffused necking and can be defined as a function of stress-state variable $\varepsilon_f(\eta,\xi)$. The undeformed material is characterized by $F = 0$, while $F = 1$ indicates the initiation of damage localization. Upon reaching $F = 1$, the current damage value is set to the critical damage value ($D_{crit}$) [24,52]. Beyond this point, the flow stress and damage are coupled as per Equation (24), where $\tilde{\sigma}$ is the undamaged stress tensor, and $m$ is the fade exponent, which governs the rate of energy dissipation and stress decay and is typically calibrated to align with experimental results [23,54]. The model also offers two alternative methods for defining the coupling between flow stress and damage. The first approach involves specifying a constant critical damage value ($D_{crit}$). In this case, coupling is initiated when the current damage value ($D$) surpasses $D_{crit}$. The second approach sets a fixed value for the critical plastic strain. Once the plastic strain reaches this predefined value, the current damage is stored as $D_{crit}$, thereby activating stress-damage coupling [49].

$$\Delta F = \frac{DMGEXP}{\varepsilon_{crit}(\eta, \xi)} F^{\left(\frac{DMGEXP-1}{DMGEXP}\right)} \Delta \varepsilon_p \tag{23}$$

$$\sigma = \tilde{\sigma}\left[1 - \left(\frac{D - D_{crit}}{1 - D_{crit}}\right)^m\right] \tag{24}$$

4.1.3. Piecewise-Linear Plasticity Model

The *Piecewise Linear Plasticity* material model (*MAT24*) in LS-DYNA 14.0 is frequently used in crashworthiness applications [1,49]. Despite the simplified failure criterion, which is a constant effective plastic strain at failure, MAT24 offers several benefits over other material models, such as its computational-cost efficiency and strain-rate dependency [35]. This material model is based on elasto-plasticity and handles strain-rate effects using the Cowper–Symonds strain-rate model, and includes the von Mises yield criterion, the associated flow rule, and isotropic strain hardening. In this model, the deviator stress is determined to satisfy the yield function as provided in Equation (25), where $s_{ij}$ are the components of the deviatoric stress tensor. Yield stress, $\sigma_y$ is expressed in Equation (26). In this equation, $\beta$ is a strain rate factor and accounts for strain-rate effects, $\sigma_0$ is the initial yield stress, $f_h\left(\varepsilon^p_{eff}\right)$ is the hardening function, and can be specified in tabular form or linear hardening of the form $f_h\left(\varepsilon^p_{eff}\right) = E_P\varepsilon^p_{eff}$ with plastic hardening modulus $E_P$.

$$f = \frac{1}{2}s_{ij}s_{ij} - \left(\frac{\sigma_y}{\sqrt{3}}\right)^2 \leq 0 \tag{25}$$

$$\sigma_y = \beta\left[\sigma_0 + f_h\left(\varepsilon^p_{eff}\right)\right] \tag{26}$$

The deviatoric stress is updated elastically, and the yield function is checked. The deviatoric stress is accepted if the yield function is satisfied. Otherwise, the plastic strain increment is calculated as given in Equation (27), where $E_P$ is the current hardening modulus, and $G$ is the shear modulus. The trial deviatoric stress state, $\tilde{s}_{ij}$ is scaled back as expressed in Equation (28).

$$\Delta\varepsilon^p_{eff} = \frac{\left(\frac{3}{2}\tilde{s}_{ij}\tilde{s}_{ij}\right)^{\frac{1}{2}} - \sigma_y}{E_P + 3G} \tag{27}$$

$$s^{n+1}_{ij} = \frac{\sigma_y}{\left(\frac{3}{2}\tilde{s}_{kl}\tilde{s}_{kl}\right)^{\frac{1}{2}}}\tilde{s}_{ij} \tag{28}$$

The Cowper–Symonds model [55] scales the yield stress using a factor $\beta$, which is calculated following Equation (29), in which, $\dot{\varepsilon}_p$ is the effective plastic strain rate, and $c$ and $p$ are Cowper–Symonds strain rate parameters.

$$\beta = 1 + \left(\frac{\dot{\varepsilon}_p}{c}\right)^{\frac{1}{p}} \tag{29}$$

*4.2. Development of Stress-State-Dependent Failure Surface*

In the context of this paper and other related studies, failure surface refers to the three-dimensional representation of failure strain as a function of stress triaxiality and Lode parameter. Extensive research has been carried out to develop stress-state-dependent failure surfaces that can be utilized in finite element modeling to simulate the fracture behavior of metals [16,18,20,25,56]. Smoothed thin-plate spline (TPS), which is a mathematical interpolation technique used to fit smooth surfaces to a scattered dataset, was employed to establish a baseline failure surface utilizing test results reported by Buyak on Al2024-T351 aluminum, assuming similar response and fracture behavior [1,25,57]. Following a series of FE simulations, and as expected, this failure surface was unable to accurately capture failure under various loading conditions. Hence, the surface was modified to fit the test results summarized previously. In these tests, each specimen configuration is attributed to a unique stress state (i.e., $\eta$ and $\xi$) and failure strain value, which is represented a single point on the failure surface. Despite being theoretically possible, obtaining every data point of stress triaxiality and Lode parameter to develop a complete failure surface is practically unachievable due to extensive and complex testing requirements. Although utilizing 21 specimens to construct the failure surface was deemed adequate compared to relevant research studies, additional numerical simulations were carried out to provide a more accurate and comprehensive representation of ductile failure [18,25,43,58]. In these simulations, test specimens were modeled, and the *MAT24* and *GISSMO* models were adopted to simulate their response. The failure surface was then tweaked to achieve a reasonable match between numerical and test results. LS-DYNA's **CONTROL_DEBUG* keyword was implemented to extract the stress-state parameters and the corresponding failure strain required to modify the failure surface for each simulated specimen. Consequently, the final optimized failure surface is depicted in Figure 9. As described earlier, the *GISSMO* model enables the users to define critical strain, which is the strain at the onset of diffused necking, as a function of stress triaxiality and Lode parameter. This secondary surface was also developed following the same approach, as illustrated in Figure 10.

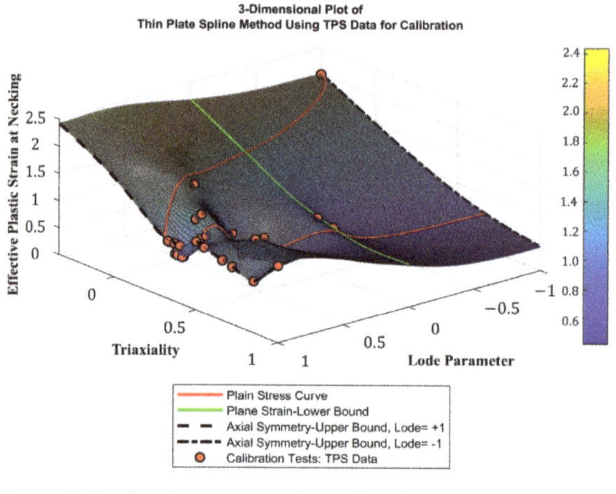


**Figure 9.** Finalized stress-state-dependent failure surface.

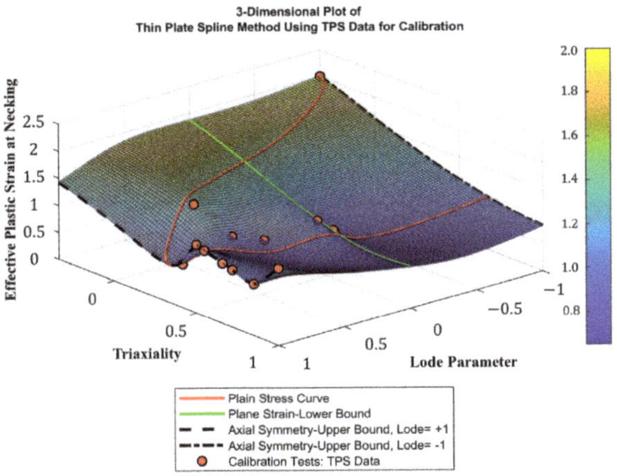


**Figure 10.** Critical stress-state-dependent surface.

### *4.3. FE Models Development*

A series of finite element simulations was carried out using LS-DYNA 14.0 to demonstrate the capability of the identified material properties and the stress-state-dependent failure surface in replicating test results. These material characteristics were used to define all inputs for the previously presented three material models (*MAT24, MAT224,* and *GISSMO*). In addition to material models, FE simulations considered various model parameters, such as element types and formulations. Comparisons between simulation and test results are presented in this section.

Tension test specimens no. 1 through no. 13 were modeled and simulated using shell and solid elements. Extensive mesh sensitivity analyses were initially completed, and more details can be found elsewhere [1,40]. According to these studies, hexahedral solid elements and quadrilateral shell elements were employed to model the test specimens. It was determined that limiting element size to 0.152 mm or less in the vicinity of the necking region would provide accurate results. Modeled specimens are illustrated in Figures 11 and 12. To examine the robustness of the developed models, under-integrated (*UI*) and fully-integrated (*FI*) quadrilateral shell elements were initially employed to model specimens no. 1 through no. 4. For the *UI* shell elements, the default Belytschko–Tsay (*ELFORM* = 2) was used. *FI* shell elements were modeled using the thickness-enhanced shell element formulation (*ELFORM* = 26) with 2 × 2 integration points. The shell thickness update option was activated, which accounts for the effects of shell membrane thinning, by setting *ISTUPD* to 1 in the *Control_Shell* keyword. To counterbalance zero energy distortion modes, the standard LS-DYNA's viscous hourglass control form (*IHQ* = 1) with an hourglass coefficient (*QH*) of 0.01. Similar to shell elements, *UI* and *FI* solid elements were utilized to model all test specimens. The default constant stress solid elements in LS-DYNA 14.0 (*ELFORM* = 1) with single integration point and type 6 hourglass control with a coefficient value of 0.01 were used. For *FI* solid elements, an improved and efficient 8-point hexahedral solid element formulation (*ELFORM* = −1) was used.

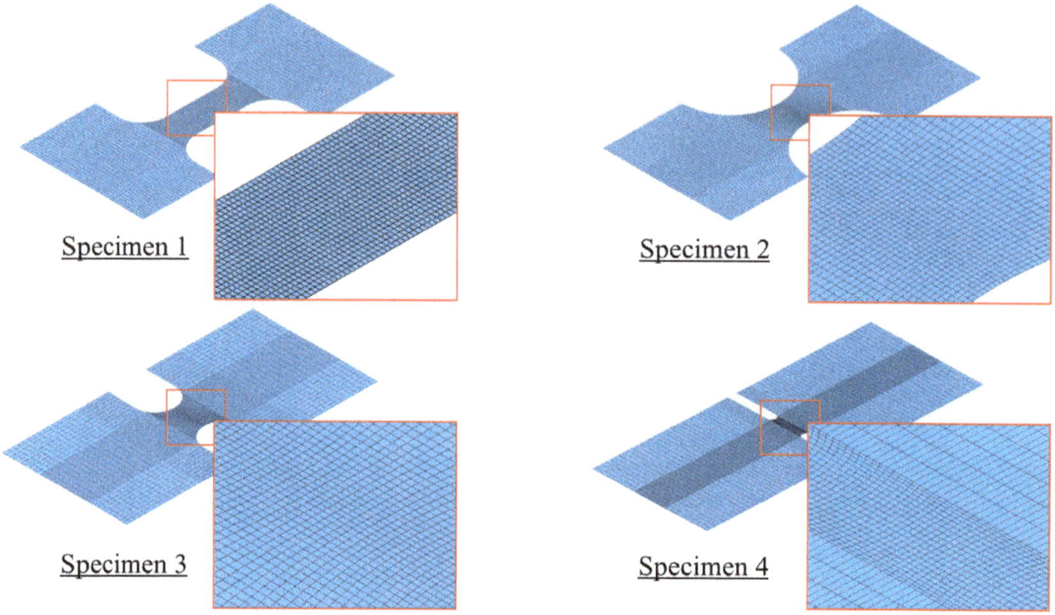


**Figure 11.** FE models of specimens 1 to 4 meshed using shell elements.

Following the model construction, the previously discussed material models, *MAT224* and *MAT24*, were used with and without *GISSMO*. In the present study, *MAT24* was examined when used individually or coupled with the *GISSMO* damage model to allow for stress-state-dependent failure criteria to be defined. Consequently, the robustness of the derived material characteristics was investigated through the implementation of four modeling scenarios: (i) *MAT224* only; (ii) *MAT224* with *GISSMO*; (iii) *MAT24* only; and (iv) *MAT24* with *GISSMO*. In the case that involved *GISSMO*, density, elastic modulus, Poisson's ratio, yield strength, and plasticity curve were defined within *MAT24* and *MAT224* models, while the failure surface and the damage parameters presented earlier were specified within *ADD_DAMAGE_GISSMO* control cards. Basic material properties and the derived plasticity curve are provided in Tables 4 and 5, respectively. When *MAT224* was solely used, the failure surface was used as an input in conjunction with the previously mentioned material parameters. For the case in which only *MAT24* was employed, the plastic failure strain (*FAIL*) was set to 1.40 to account for element erosion.

**Table 4.** Basic input material parameters.

| Parameter | Value |
|---|---|
| Mass density (kg/m$^3$) | 7850 |
| Elastic modulus (MPa) | $1.993 \times 10^5$ |
| Yield strength (MPa) | 439.6 |
| Poisson's ratio | 0.295 |
| Plastic failure strain | 1.400 |

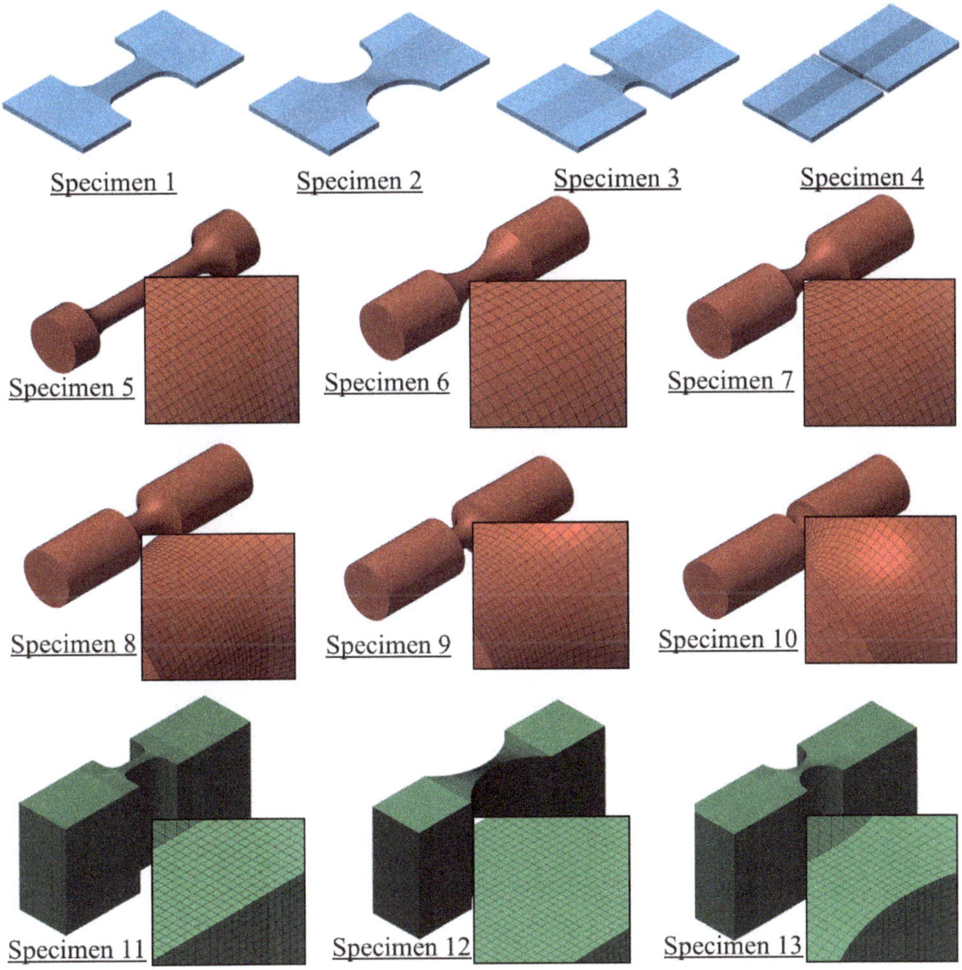


**Figure 12.** FE models of specimens 1 to 13 meshed using solid elements.

**Table 5.** Input plasticity curve.

| Effective Plastic Strain | Effective Stress (GPa) |
|---|---|
| 0.000 | 0.4396 |
| 0.005 | 0.4634 |
| 0.025 | 0.5174 |
| 0.040 | 0.5419 |
| 0.075 | 0.5807 |
| 0.145 | 0.6278 |
| 0.300 | 0.6870 |
| 0.500 | 0.7329 |
| 0.700 | 0.7652 |
| 0.900 | 0.7903 |
| 1.000 | 0.8010 |
| 1.200 | 0.8201 |
| 1.400 | 0.8365 |

## 5. FE Simulation Results

In this section, simulated and actual engineering stress–strain curves were compared for all modeled test specimens, as shown in Figures 13–17. In Figures 13 and 14, each curve represents a particular material modeling scenario when shell and solid elements were utilized. Again, all material models were explored using *UI* and *FI* element formulations. Irrespective of the material model, it was noted that specimens modeled using *UI* shell elements were not able to accurately reproduce test results. On the other hand, the *FI* shell elements demonstrated good agreement with test results for specimens 1 to 3. However, both *UI* and *FI* shell elements showed a noticeable deviation from the test results of specimen 4. This can be attributed to the complex stress state that this particular specimen configuration encompasses. As described earlier, specimen no. 4 corresponds to a plane strain loading condition characterized by $\xi = 0$, which requires a three-dimensional element representation to account for out-of-plane stresses. This was confirmed when a better correlation between the actual and simulated curves was achieved when solid elements were used, with no significant variation between *UI* and *FI* element formulations. Furthermore, it was obvious that using *GISSMO* with the selected material models enhances the prediction capability of the developed FE models, specifically, under complex loading conditions.

Results from specimens no. 5 to no. 10 are illustrated in Figures 15 and 16. Again, integrating the *GISSMO* damage model with *MAT24* contributed to better results, demonstrating enhanced prediction capability. On the other hand, *MAT224* performed largely the same with and without coupling it with *GISSMO*. However, any slight discrepancy between the modeling approaches can be attributed to the coupling between flow stress and damage uniquely supported by *GISSMO*. For any given material modeling scenario, the response of specimens modeled using *UI* and *FI* solid elements demonstrated good agreement. Accordingly, *UI* solid elements are recommended given their considerably lower computational cost.

Similarly, the proposed material modeling approach is deemed efficient for the response of the more complex specimens (i.e., nos. 11–13), as highlighted in Figure 17. Following the previous observations, incorporating the *GISSMO* damage model into *MAT24* and *MAT224* resulted in improved response and failure prediction, with a more prominent enhancement when *MAT24* was implemented. Particularly, it was observed that using the *GISSMO* damage model improved the capability of the FE models in predicting the most complex loading scenarios attributed to specimens no. 11 through no. 13. Although the ultimate strength is slightly overestimated in specimens no. 11 and no. 12, the overall response of these test specimens is precisely reproduced. As this study intends to avoid a tedious model refinement process by providing a simplified modeling approach and all necessary material model parameters needed to readily simulate the behavior of guardrail steel, no further model tweaking was performed.

To further evaluate the proposed material modeling approach, failure strains predicted by the FE simulations were compared to those obtained from testing, with simulated failure strains referring to the first failed element. Each strain value was then normalized to the corresponding actual failure strain, as shown in Figure 18. This figure emphasizes the accuracy of the derived input parameters and the capability of the selected modeling scenarios to predict the actual failure strain. For solid elements depicted in Figure 18a, it was emphasized that *MAT24* and *MAT224* material models experienced better prediction accuracy when coupled with *GISSMO*. Generally, using *MAT24* only resulted in overestimated failure strain values. On the other hand, using *MAT224* is attributed to relatively better predictions. This is mainly because *MAT224* utilizes stress-state-dependent failure surface to define ductile fracture, which is largely the same approach available in the *GISSMO* damage model. It was also noted that *FI* and *UI* solid element formulations largely behaved similarly. For shell elements illustrated by Figure 18b, only *FI* element formulations were considered in this evaluation, as they demonstrated superior performance compared to *UI* elements. Furthermore, specimen 4 was excluded from this assessment due to the poor performance associated with using shell elements. Again, *MAT24* and *MAT224* integrated with *GISSMO* experienced superior performance.

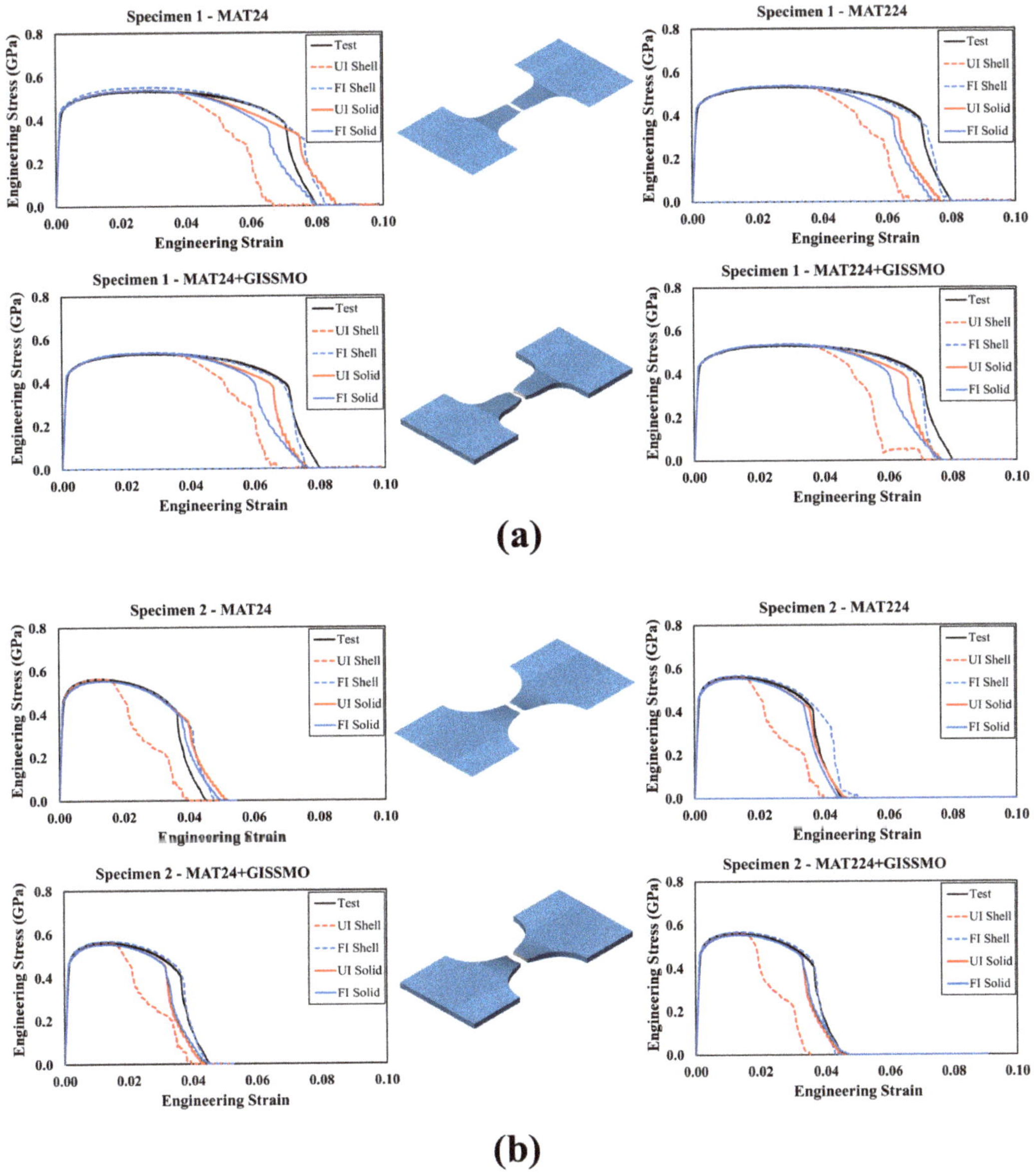


**Figure 13.** Simulated and tested engineering stress–strain curves: (**a**) specimen 1; (**b**) specimen 2.

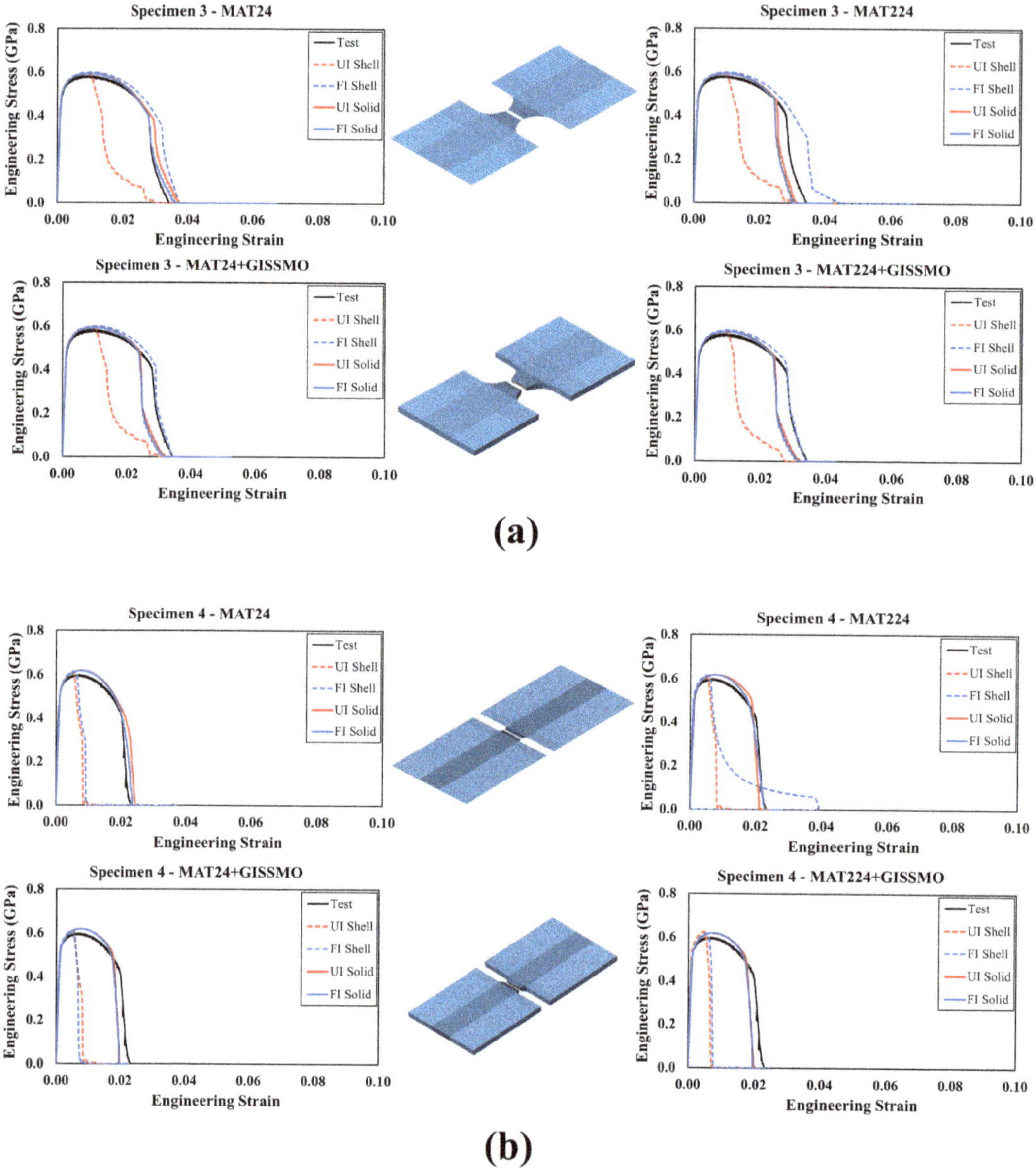


**Figure 14.** Simulated and tested engineering stress–strain curves: (**a**) specimen 3; (**b**) specimen 4.

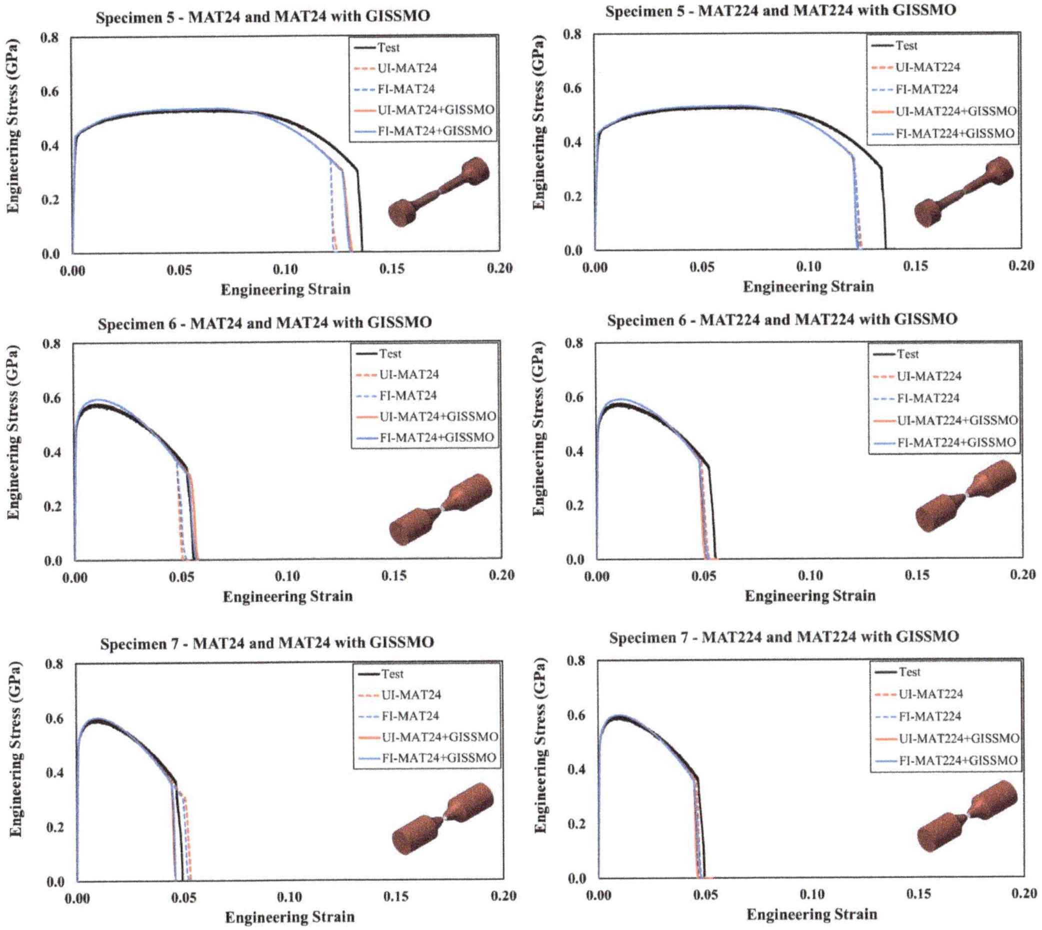


**Figure 15.** Simulated and tested engineering stress–strain curves: specimens 5 to 7.

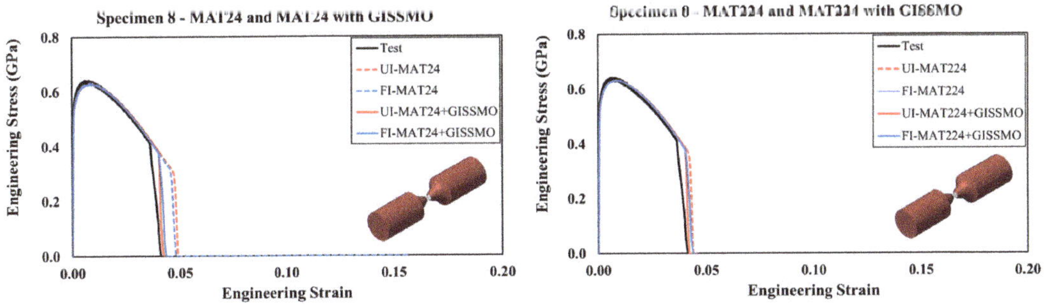


**Figure 16.** *Cont.*

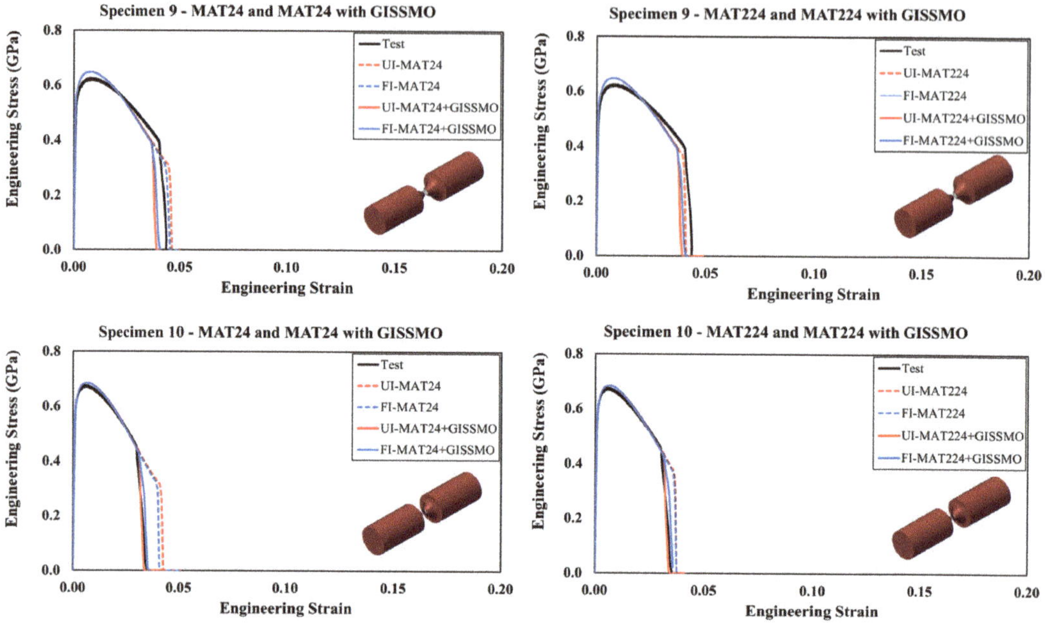


**Figure 16.** Simulated and tested engineering stress–strain curves: specimens 8 to 10.

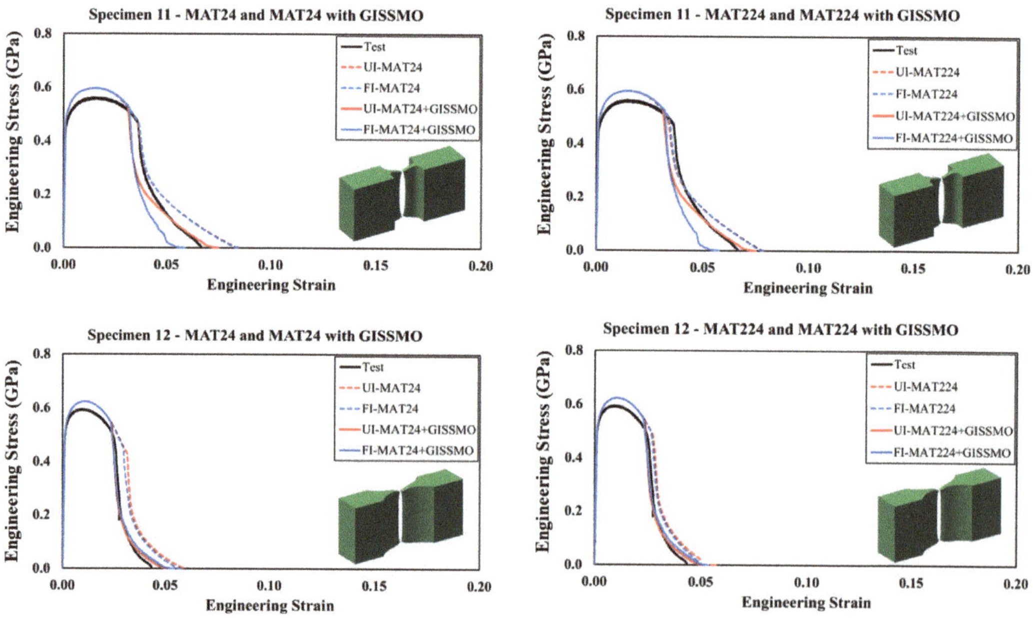


**Figure 17.** *Cont.*

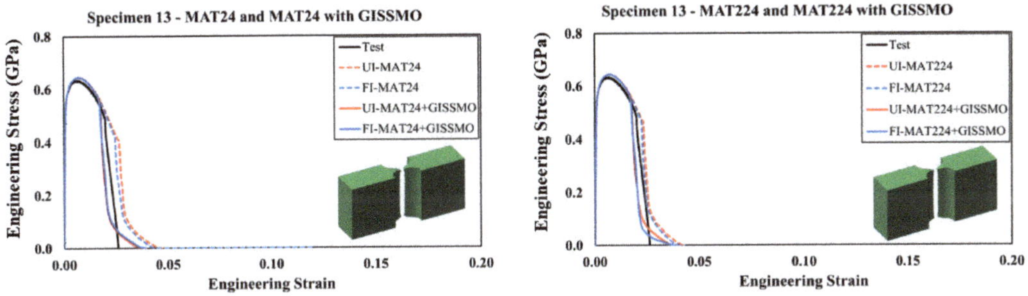


**Figure 17.** Simulated and tested engineering stress–strain curves: specimens 11 to 13.

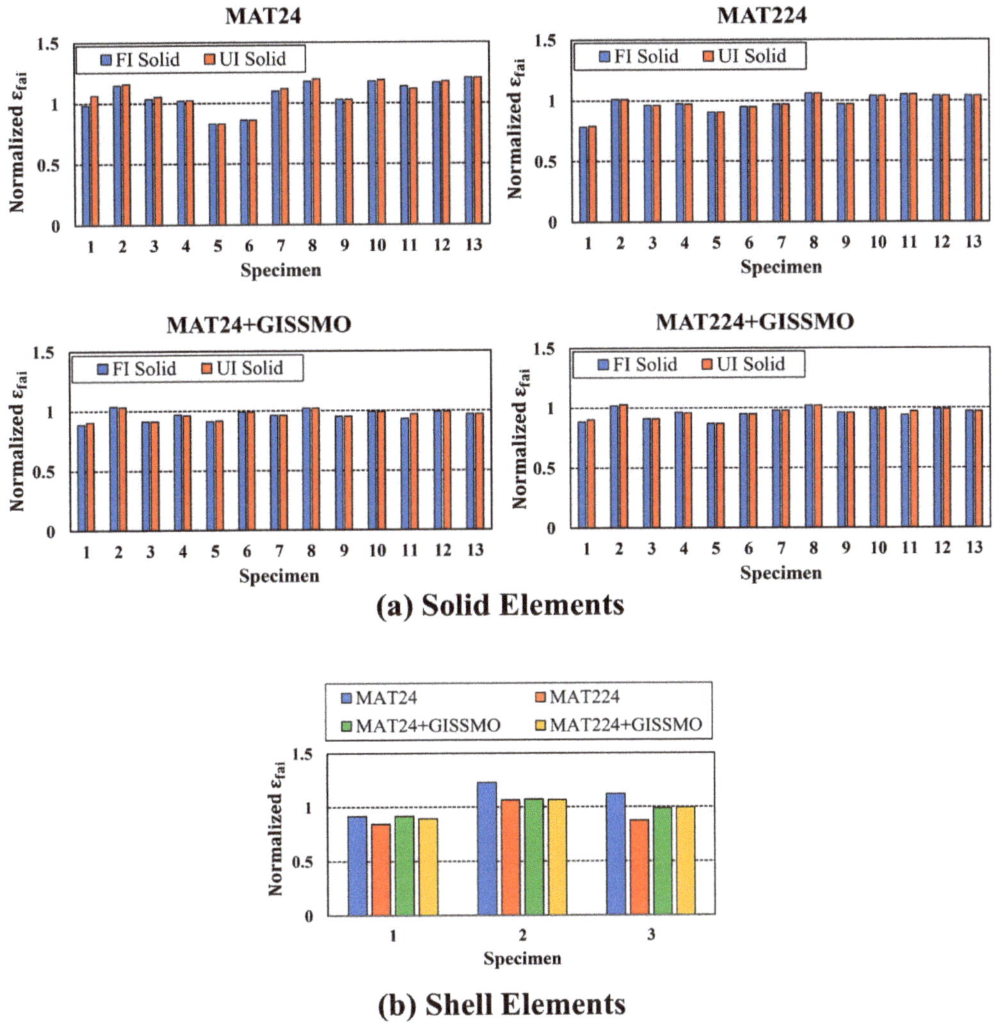


**Figure 18.** Normalized failure strain, all specimen configurations and material models: (**a**) solid elements; (**b**) shell elements.

## 6. Summary and Conclusions

This study presents a comprehensive framework for developing stress-state-dependent failure criteria and calibrated material models aimed at improving finite element predictions of ductile fracture in AASHTO M-180 guardrail steel. The work represents a significant advancement in modeling guardrail behavior for roadside safety applications, particularly through the integration of damage mechanics into LS-DYNA 14.0 simulations.

Due to fabrication limitations associated with W-beam sections, experimental testing was conducted using ASTM A572 Grade 50 steel, selected for its mechanical similarity to AASHTO M-180. Twenty-one uniquely designed specimens were tested under various loading conditions to span a broad range of stress triaxiality ($\eta$) and Lode parameter ($\xi$), enabling the development of a stress-state-dependent failure surface. This surface relates equivalent plastic failure strain ($\varepsilon^P{}_f$) to $\eta$ and $\xi$, forming the foundation for material input calibration.

The extracted material parameters were implemented into LS-DYNA 14.0 using *MAT24* and *MAT224* models, with and without the Generalized Incremental Stress-State-Dependent Model (*GISSMO*). A comprehensive simulation study was performed using different element types and formulations, and the results were compared to experimental data. The following conclusions were drawn:

1. The developed failure surface and material parameters allow accurate simulation of guardrail steel fracture under complex loading conditions.
2. *GISSMO* played a critical role in improving prediction accuracy, particularly under multiaxial stress states, and its coupling with *MAT24* and *MAT224* resulted in significantly better agreement with experimental results.
3. While both *MAT224* and *GISSMO* can implement the failure surface, slight differences in response are due to *GISSMO*'s ability to couple damage evolution with flow stress, which *MAT224* lacks.
4. Fully integrated shell elements (*FI*) outperformed under-integrated (*UI*) ones in capturing experimental trends, though shell elements alone were insufficient for capturing complex stress states.
5. Solid elements (*UI* and *FI*) proved more appropriate in multiaxial loading scenarios, with UI solids recommended for their balance of accuracy and computational efficiency.

Importantly, the study provides a validated modeling framework and recommended LS-DYNA 14.0 inputs that can be directly and efficiently implemented in future roadside safety simulations. All inputs were derived from physical testing and constructed without arbitrary fitting, ensuring both transparency and practical utility. Ongoing research will focus on mesh regularization to ensure solution consistency across varying element sizes, a common challenge in FE modeling. In parallel, validation against full-scale crash testing will be pursued to further strengthen the model's applicability to real-world roadside safety design and evaluation.

**Author Contributions:** Study conception and design: Q.A.A., R.W.B., R.K.F., M.N., Z.Z. and W.L.; data collection: Q.A.A., Z.Z., W.L. and B.M.H.; analysis and interpretation of results: Q.A.A., R.W.B. and R.K.F.; draft manuscript preparation: Q.A.A., T.Y.Y. and M.N. All authors have read and agreed to the published version of the manuscript.

**Funding:** This research was funded in part by Earmark support (State PN 00965), which covered the initial phase of the project, and in part by the Midwest Pool Fund Program (POOLFUND130220), which supported the simulation efforts.

**Institutional Review Board Statement:** Not applicable.

**Informed Consent Statement:** Not applicable.

**Data Availability Statement:** Data is available on request due to restrictions. The data presented in this study are available on request from the corresponding author. The data is not publicly available because this study is part of an ongoing large research project.

**Acknowledgments:** The authors wish to acknowledge the Midwest Pooled Fund program for funding these research efforts. This work was completed utilizing the Holland Computing Center of the University of Nebraska, which receives support from the Nebraska Research Initiative.

**Conflicts of Interest:** Author Zesheng Zhang was employed by the company Dimensional Control Systems. Author Brandt M. Humphrey was employed by the company Burns and McDonnell Construction. The remaining authors declare that the research was conducted in the absence of any commercial or financial relationships that could be construed as a potential conflict of interest.

## List of Abbreviations

Note: Other variables are unique to specific material models in LS-DYNA 14.0 and are defined where they first appear in the text.

| Abbreviation | Definition |
|---|---|
| AASHTO | American Association of State Highway and Transportation Officials |
| ASTM | American Society for Testing and Materials |
| MASH | Manual of Assessing Safety Hardware |
| FEA | Finite Element Analysis |
| LS-DYNA | Livermore Software Technology Corporation—Dynamic Nonlinear Analysis (FEA Software) |
| $\sigma$ | Cauchy stress tensor |
| $\sigma_H$ | Hydrostatic stress |
| $s$ | Deviatoric stress |
| $tr(\sigma)$ | Trace of the stress tensor, which is the sum of its diagonal components |
| $\boldsymbol{I}$ | The second-order identity tensor |
| $P$ | Pressure |
| $I_1, I_2, I_3$ | Stress invariants |
| $\mathbf{s}:\mathbf{s}$ | double contraction of the deviatoric stress tensor with itself |
| $\sigma_m$ | Von Mises or average stress |
| $\eta$ | Stress triaxiality |
| $\xi$ | Lode parameter |
| $J_3$ | Third deviatoric stress invariant |
| DIC | Digital Image Correlation |
| $\sigma_T$ | True uniaxial stress |
| $\varepsilon_T$ | True uniaxial strain |
| $K$, $\varepsilon_o$, and $n$ | Swift plasticity model constants |
| MAT224 | Tabulated Johnson–Cook LS-DYNA's material model |
| MAT24 | Piecewise Linear Plasticity LS-DYNA's material model |
| GISSMO | Generalized Incremental Stress-State-Dependent Model |
| ELFORM | Element formulation in LS-DYNA |
| FI | Fully Integrated elements |
| UI | Under Integrated elements |

## References

1. Humphrey, B.M.; Faller, R.K.; Bielenberg, R.W.; Reid, J.D. *Improved Methodologies in Modeling and Predicting Failure in AASHTO M-180 Guardrail Steel Using Finite Element Analysis-Phase I*; Midwest Roadside Safety Facility (MwRSF), University of Nebraska–Lincoln: Lincoln, NE, USA, 2016.
2. American Association of State Highway and Transportation Officials (AASHTO). *Manual for Assessing Safety Hardware (MASH)*, 2nd ed.; American Association of State Highway and Transportation Officials (AASHTO): Washington, DC, USA, 2016.
3. Gabauer, D.J.; Kusano, K.D.; Marzougui, D.; Opiela, K.; Hargrave, M.; Gabler, H.C. Pendulum testing as a means of assessing the crash performance of longitudinal barrier with minor damage. *Int. J. Impact Eng.* **2010**, *37*, 1121–1137. [CrossRef]

4. Mohr, D.; Henn, S. Calibration of Stress-triaxiality Dependent Crack Formation Criteria: A New Hybrid Experimental–Numerical Method. *Exp. Mech.* **2007**, *47*, 805–820. [CrossRef]
5. Winkelbauer, B.J.; Putjenter, J.G.; Rosenbaugh, S.K.; Lechtenberg, K.A.; Bielenberg, R.W.; Faller, R.K.; Reid, J.D. *Dynamic Evaluation of MGS Stiffness Transition with Curb*; Midwest Roadside Safety Facility: Lincoln, NE, USA, 2014.
6. Bielenberg, R.W.; Ahlers, T.J.; Faller, R.K.; Holloway, J.C. MASH Testing of Bullnose with Breakaway Steel Posts (Test Nos. MSPBN-4 Through MSPBN-8). 2020. Available online: https://mwrsf.unl.edu/researchhub/files/Report393/TRP-03-418-20.pdf (accessed on 23 May 2025).
7. Basaran, M.; Wölkerling, S.D.; Feucht, M.; Neukamm, F.; Weichert, D.; DAIMLER AG, Sindelfingen. An extension of the GISSMO damage model based on lode angle dependence. *LS-DYNA Anwenderforum* **2010**, *15*, 15.
8. Kim, J.-H.; Serpantié, A.; Barlat, F.; Pierron, F.; Lee, M.-G. Characterization of the post-necking strain hardening behavior using the virtual fields method. *Int. J. Solids Struct.* **2013**, *50*, 3829–3842. [CrossRef]
9. *AASHTO M 180-12*; Standard Specification for Corrugated Sheet Steel Beams of Highway Gaurdrail. AASHTO Designation: Washington, DC, USA, 2012.
10. Yu, M. Advances in strength theories for materials under complex stress state in the 20th century. *Appl. Mech. Rev.* **2002**, *55*, 169–218. [CrossRef]
11. Won, J.Y.; Kim, C.; Hong, S.; Yoon, H.-S.; Park, J.K.; Lee, M.-G. Evaluation of crush performance of extruded aluminum alloy tubes based on finite element analysis with ductile fracture modeling. *Thin-Walled Struct.* **2024**, *200*, 111937. [CrossRef]
12. Lou, Y.; Chen, L.; Clausmeyer, T.; Tekkaya, A.E.; Yoon, J.W. Modeling of ductile fracture from shear to balanced biaxial tension for sheet metals. *Int. J. Solids Struct.* **2017**, *112*, 169–184. [CrossRef]
13. Yang, X.; Guo, Y.; Li, Y. A new ductile failure criterion with stress triaxiality and Lode dependence. *Eng. Fract. Mech.* **2023**, *289*, 109394. [CrossRef]
14. Bridgman, P.W. *Studies in Large Plastic Flow and Fracture: With Special Emphasis on the Effects of Hydrostatic Pressure*; Harvard University Press: Cambridge, MA, USA, 1964. [CrossRef]
15. Bao, Y. Prediction of Ductile Crack Formation in Uncracked Bodies. Ph.D. Thesis, Massachusetts Institute of Technology, Cambridge, MA, USA, 2003.
16. Wierzbicki, T.; Bao, Y.; Lee, Y.-W.; Bai, Y. Calibration and evaluation of seven fracture models. *Int. J. Mech. Sci.* **2005**, *47*, 719–743. [CrossRef]
17. Carney, K.S.; DuBois, P.A.; Buyuk, M.; Kan, S. Generalized, Three-Dimensional Definition, Description, and Derived Limits of the Triaxial Failure of Metals. *J. Aerosp. Eng.* **2009**, *22*, 280–286. [CrossRef]
18. Bai, Y.; Wierzbicki, T. A new model of metal plasticity and fracture with pressure and Lode dependence. *Int. J. Plast.* **2008**, *24*, 1071–1096. [CrossRef]
19. Jia, L.-J.; Ge, H.; Shinohara, K.; Kato, H. Experimental and Numerical Study on Ductile Fracture of Structural Steels under Combined Shear and Tension. *J. Bridge Eng.* **2016**, *21*, 04016008. [CrossRef]
20. Bai, Y.; Teng, X.; Wierzbicki, T. On the application of stress triaxiality formula for plane strain fracture testing. *J. Eng. Mater. Technol.* **2009**, *131*, 021002. [CrossRef]
21. Landes, J.D.; McCabe, D.E.; Boulet, J.A.M. *Fracture Mechanics: Twenty-Fourth Volume*; ASTM International: West Conshohocken, PA, USA, 1994; Volume 1207.
22. Ganjiani, M. A damage model for predicting ductile fracture with considering the dependency on stress triaxiality and Lode angle. *Eur. J. Mech.-A/Solids* **2020**, *84*, 104048. [CrossRef]
23. Topilla, L.; Toros, S. Analysis of the Fracture Behaviour of Dual-phase Steels Using the GISSMO and Johnson Cook Models. *Trans. FAMENA* **2023**, *47*, 79–95. [CrossRef]
24. Andrade, F.X.C.; Feucht, M.; Haufe, A.; Neukamm, F. An incremental stress state dependent damage model for ductile failure prediction. *Int. J. Fract.* **2016**, *200*, 127–150. [CrossRef]
25. Buyuk, M. *Development of a Tabulated Thermo-Viscoplastic Material Model with Regularized Failure for Dynamic Ductile Failure Prediction of Structures Under Impact Loading*; The George Washington University: Washington, DC, USA, 2013.
26. Gao, X.; Zhang, G.; Roe, C. A Study on the Effect of the Stress State on Ductile Fracture. *Int. J. Damage Mech.* **2010**, *19*, 75–94. [CrossRef]
27. Kiran, R.; Khandelwal, K. Experimental Studies and Models for Ductile Fracture in ASTM A992 Steels at High Triaxiality. *J. Struct. Eng.* **2014**, *140*, 04013044. [CrossRef]
28. Shen, F.; Pan, B.; Wang, S.; Lian, J.; Münstermann, S. Influence of stress states on cleavage fracture in X70 pipeline steels. *J. Pipeline Sci. Eng.* **2022**, *2*, 100072. [CrossRef]
29. Brünig, M.; Schmidt, M.; Gerke, S. A fracture criterion for ductile metals based on critical damage parameters: Dedicated to Professor Holm Altenbach on the occasion of his 65th birthday. *Contin. Mech. Thermodyn.* **2021**, *33*, 1011–1022. [CrossRef]
30. Brünig, M.; Schmidt, M.; Gerke, S. Numerical analysis of stress-state-dependent damage and failure behavior of ductile steel based on biaxial experiments. *Comput. Mech.* **2021**, *68*, 1–11. [CrossRef]

31. Zhang, Z.; Shen, F.; Liu, H.; Könemann, M.; Münstermann, S. Temperature-dependent deformation and fracture properties of low-carbon martensitic steel in different stress states. *J. Mater. Res. Technol.* **2023**, *25*, 1931–1943. [CrossRef]
32. Testa, G.; Ricci, S.; Iannitti, G.; Ruggiero, A.; Bonora, N. Modeling ductile fracture in third stress invariant sensitive materials: Application to Al 2024-T351. *Mech. Mater.* **2023**, *179*, 104584. [CrossRef]
33. Brünig, M.; Zistl, M.; Gerke, S. Biaxial experiments on characterization of stress-state-dependent damage in ductile metals. *Prod. Eng. Res. Devel.* **2020**, *14*, 87–93. [CrossRef]
34. Brünig, M.; Koirala, S.; Gerke, S. A stress-state-dependent damage criterion for metals with plastic anisotropy. *Int. J. Damage Mech.* **2023**, *32*, 811–832. [CrossRef]
35. Svanberg, A. Numerical Methods for Simulating the Metal Shearing Process: A Novel Numerical Model for the Punching of Metals. Master's Thesis, Luleå University of Technology, Luleå, Sweden, 2019.
36. Irgens, F. *Continuum Mechanics*; Springer Science & Business Media: New York, NY, USA, 2008.
37. Fung, Y.C.; Tong, P.; Chen, X.H. *Classical and Computational Solid Mechanics*, 2nd ed.; World Scientific: Singapore, 2017; Volume 2, ISBN 978-981-4713-64-1.
38. Huang, Y.; Jiang, J. A critical review of von mises criterion for compatible deformation of polycrystalline materials. *Crystals* **2023**, *13*, 244. [CrossRef]
39. Schmidt, J.D.; Bielenberg, R.W.; Reid, J.D.; Faller, R.K. *Numerical Investigation on the Performance of Steel Guardrail Systems with Varied Mechanical Properties*; Midwest Roadside Safety Facility, University of Nebraska-Lincoln: Lincoln, NE, USA, 2013.
40. Schrum, K.D.; Sicking, D.L.; Faller, R.K.; Reid, J.D. *Predicting the Dynamic Fracture of Steel via a Non-Local Strain-Energy Density Failure Criterion*; The University of Nebraska-Lincoln: Lincoln, NE, USA, 2014.
41. *ASTM A572-15*; Standard Specification for High-Strength Low-Alloy Columbium-Vanadium Structural Steel. ASTM: West Conshohocken, PA, USA, 2015.
42. Mirone, G. Role of stress triaxiality in elastoplastic characterization and ductile failure prediction. *Eng. Fract. Mech.* **2007**, *74*, 1203–1221. [CrossRef]
43. Seidt, J.D. *Development of a New Metal Material Model in LS-DYNA Part 3: Plastic Deformation and Ductile Fracture of 2024 Aluminum Under Various Loading Conditions*; Finnal Report, DOT/FAA/TC-13/25; Federal Aviation Administration (FAA): Washington, DC, USA, 2014; p. 3.
44. Faridmehr, I.; Osman, M.H.; Adnan, A.B.; Nejad, A.F.; Hodjati, R.; Azimi, M. Correlation between engineering stress-strain and true stress-strain curve. *Am. J. Civ. Eng. Archit.* **2014**, *2*, 53–59. [CrossRef]
45. Bhonge, P.S.; Lankarani, H.M. Fine-tuning nonlinear finite element analysis methodology for aircraft seat certification using component level testing and validation. *Int. J. Veh. Struct. Syst.* **2011**, *3*, 129–138. [CrossRef]
46. Paul, S.K.; Roy, S.; Sivaprasad, S.; Bar, H.N.; Tarafder, S. Identification of Post-necking Tensile Stress–Strain Behavior of Steel Sheet: An Experimental Investigation Using Digital Image Correlation Technique. *J. Mater. Eng Perform* **2018**, *27*, 5736–5743. [CrossRef]
47. Swift, H. Plastic instability under plane stress. *J. Mech. Phys. Solids* **1952**, *1*, 1–18. [CrossRef]
48. Vasko, T. A User Guide for *MAT_TABULATED_JOHNSON_COOK in LS-DYNA. LS-DYNA Aerospace Working Group. 2021. Available online: https://awg.ansys.com/tiki-download_file.php?fileId=2001 (accessed on 23 May 2025).
49. Hallquist, J.O. *LS-DYNA® Keyword User's Manual: Volumes I, II, and III LSDYNA R7*; 1. Livermore Software Technology Corporation, Livermore (LSTC): Livermore, CA, USA, 2014; p. 1265.
50. Andrade, F.; Feucht, M.; Haufe, A. On the Prediction of Material Failure in LS-DYNA®: A Comparison Between GISSMO and DIEM. 2014. Available online: https://lsdyna.ansys.com/wp-content/uploads/2025/02/on-the-prediction-of-material-failure-in-ls-dyna-a-comparison-between-gissmo-and-diem.pdf (accessed on 23 May 2025).
51. Neukamm, F.; Feucht, M.; Haufe, A. Considering Damage History in Crashworthiness Simulations. In Proceedings of the 7th European LS-DYNA Conference, Salzburg, Austria, 14–15 May 2009.
52. Haufe, A.; Neukamm, F.; Feucht, M.; DuBois, P.; Borvall, T. A comparison of recent damage and failure models for steel materials in crashworthiness application in LS-DYNA. In Proceedings of the 11th International LS-DYNA Users Conference, Dearborn, MI, USA, 6–8 June 2010; pp. 6–8.
53. Han, F.; Lian, C.; Zhang, J.; Jiang, H. Plasticity and Fracture Characterization of Advanced High-Strength Steel: Experiments and Modeling. In *Forming the Future*; Daehn, G., Cao, J., Kinsey, B., Tekkaya, E., Vivek, A., Yoshida, Y., Eds.; Springer International Publishing: Cham, Switzerland, 2021; pp. 1947–1959. ISBN 978-3-030-75380-1.
54. Effelsberg, J.; Haufe, A.; Feucht, M.; Neukamm, F.; Du Bois, P. On parameter identification for the GISSMO damage model. In Proceedings of the 12th International LS-DYNA Users Conference, Metal Forming, Detroit, MI, USA, 3–5 June 2012; pp. 1–12.
55. Cowper, G.R.; Symonds, P.S. *Strain-Hardening and Strain-Rate Effects in the Impact Loading of Cantilever Beams*; Brown University: Providence, RI, USA, 1957.
56. Wierzbicki, T.; Xue, L. *On the Effect of the Third Invariant of the Stress Deviator on Ductile Fracture*; Technical Report; Impact and Crashworthiness Laboratory: Cambridge, MA, USA, 2005; p. 136.

57. Stals, L.; Roberts, S. Smoothing large data sets using discrete thin plate splines. *Comput. Vis. Sci.* **2006**, *9*, 185–195. [CrossRef]
58. Bao, X.; Li, B. Residual strength of blast damaged reinforced concrete columns. *Int. J. Impact Eng.* **2010**, *37*, 295–308. [CrossRef]

*materials*

Article

# Grain Refinement and Microstructural Evolution in Cobalt-Saving 18Ni (300) Maraging Steel via Cold Deformation-Cyclic Solution Treatment

**Feng Huang [1,2], Zhe Cheng [1], Defa Li [3,*], Wei Zhang [3] and Zhili Hu [1,*]**

1 Hubei Key Laboratory of Advanced Technology for Automotive Components, Wuhan University of Technology, Wuhan 430070, China; huangfeng@whut.edu.cn (F.H.); 15802739678@163.com (Z.C.)
2 Hubei Longzhong Laboratory, Xiangyang 441022, China
3 School of Mechanical and Vehicle Engineering, West Anhui University, Lu'an 237012, China; 04000144@wxc.edu.cn
* Correspondence: 04000157@wxc.edu.cn (D.L.); zhilihu@whut.edu.cn (Z.H.)

**Abstract**

To solve the problem of inadequate plasticity of traditional processing routes in improving the plasticity of novel Co-saving 18Ni (300) maraging steel, a cold deformation-cycle solution treatment process was developed. Through systematic characterization and tensile property testing, the study focuses on elucidating the impact of the number of solution treatments on the microstructure and mechanical behavior. The results showed that with a 30% cold deformation, three times of solution treatment at 860 °C for 10 min refined the original austenite grains (equivalent circle radius: 3.3 μm) and martensite structure (length and width: 7 μm and 1.3 μm, respectively) to the utmost extent. The grains became uniformly equiaxed, and the texture was eliminated, and a moderate content (4.5%) of retained austenite was formed. At this time, the material achieves the best match between strength (tensile strength of 1240 MPa) and plasticity (elongation of 9.93%), which are increased by 15.3% and 94.3%, respectively, compared with the traditional process. Mechanistic analysis revealed that grain refinement and uniform equiaxialization were the primary drivers for enhancing strength and plasticity. This study has demonstrated that the cold deformation-cyclic solution treatment process is an effective methodology for tailoring the microstructure and mechanical properties of maraging steel.

**Keywords:** maraging steel; cold deformation; cyclic solution treatment; strengthening and toughening mechanism

## 1. Introduction

Maraging steel, a novel ultra-high-strength steel, exhibits an exceptionally low carbon content (typically <0.01%) compared to conventional steel. This characteristic distinguishes its strengthening mechanism from traditional steels: rather than relying on carbides, it achieves reinforcement through precipitates [1–3]. Nickel (Ni) serves as the primary constituent of these precipitates. During aging, Ni precipitates from the matrix and combines with titanium (Ti), molybdenum (Mo), and aluminum (Al) to form numerous fine strengthening phases (e.g., η-$Ni_3Ti$, $Ni_3Mo$, and β-NiAl). These phases act as barriers to impede dislocation motion, thereby enhancing the matrix strength via an obstacle-hardening mechanism.

The commercially prevalent maraging steel is the cobalt (Co)-containing 18Ni grade. Cobalt addition not only enhances the steel's strength but also improves its plasticity.

This dual benefit arises from two mechanisms: first, increasing Co content elevates the martensite start temperature (Ms), ensuring complete austenite-to-martensite transformation during cooling; second, Co reduces the solid solubility of alloying elements, retards substructure recovery in martensite, and promotes denser distribution of precipitated phases. Additionally, Co effectively refines grain size [4,5]. However, due to Co's scarcity and high cost, efforts have shifted toward developing Co-free or Co-saving maraging steels, with performance improvements achieved through process optimization.

Current research on Co-saving 18Ni maraging steel predominantly focuses on optimizing heat treatment processes to achieve enhanced strength-plasticity coordination, with a particular emphasis on improving plasticity. Grain refinement remains the most effective strategy to simultaneously enhance both the strength and plasticity of metallic materials. Studies have shown that when the grain size is refined to a significant extent, the strength and plasticity of the material tend to reach a plateau [6]. Therefore, achieving an optimal grain size through precise process control is essential for further improving the mechanical properties of maraging steel, as it balances the synergistic effects of strengthening mechanisms and plastic deformation capacity.

Grain refinement in metallic materials can be achieved through various approaches, including alloying element addition, cold deformation, and heat treatment [7–9]. Cold deformation processing disrupts the original austenite grains [10], effectively refining the martensitic microstructure via strain-induced fragmentation [11]. However, in traditional maraging steels, a mere increase in cold deformation degree often leads to diminishing returns in grain refinement and may induce abnormal microstructures (e.g., uneven deformation or crystalline texture), thereby deteriorating mechanical properties. Cyclic solution treatment has been demonstrated to enable further grain refinement and optimize microstructural morphology [12]. Notably, existing studies have predominantly focused on maraging steels without prior cold deformation [13–15], leaving a critical research gap regarding the synergistic effect of cold deformation and cyclic solution treatment on grain refinement and microstructural homogeneity.

Therefore, this study focuses on Co-saving 18Ni steel and addresses the microstructural defects induced by cold deformation by integrating cold deformation with cyclic solution treatment. Building on the author's previous research [16], this work emphasizes experimental investigations into the effect of the number of solution treatments. The objectives are to: (1) elucidate how cyclic solution treatment influences cold-deformed microstructures, (2) decipher the microstructural evolution mechanisms and strength-ductility synergistic principles during the cold deformation-cyclic solution treatment process, and (3) establish ideal microstructural and mechanical property foundations for subsequent aging treatment while enabling potential for further Co content reduction in maraging steel.

## 2. Material and Methods

### 2.1. Materials and Processing

In this experiment, the original cast billet of Co-saving 18Ni (300) maraging steel from a certain factory was used as the test material. The main chemical compositions are shown in Table 1. Approximately 15 kg of the billet was forged and then cut into several sheets with dimensions of 200 mm × 50 mm × 5 mm to facilitate cold deformation and cyclic solution treatment tests. All specimens were sanded with sandpaper before being heat-treated.

**Table 1.** Chemical compositions of the Co-saving 18Ni (300) test steel (wt.%).

| Composition | Ni | Co | Mo | Ti | Al | Fe |
|---|---|---|---|---|---|---|
| Content | 18.13 | 10.39 | 4.47 | 0.66 | 0.10 | Bal. |

According to the process flow diagram in Figure 1, all steel plates first underwent a vacuum annealing pretreatment of 1100 °C for 20 min, followed by furnace cooling to room temperature, to eliminate residual stresses in the material after forging. The equipment utilized was a vacuum atmosphere furnace with an operating voltage of 380 V, rated power of 12 kW, maximum temperature of 1550 °C, and argon (purity: 99.9%) as the protective gas. Heating was achieved via resistance wire induction at a rate of approximately 5 °C/min. Then, cold deformation was conducted using an N-100T variable-cross-section cold rolling mill (Huangshi Huali Metalforming Machine Tool Co., LTD, Huangshi, China) with a nominal force of 1000 kN. The steel plates were repeatedly rolled to achieve the preset deformation of 30% (from 5 mm to 3.5 mm). Finally, cyclic solution treatment was performed using a box-type resistance furnace with an operating voltage of 220 V, a rated power of 4 kW, and a maximum temperature of 1000 °C. The furnace employed resistance wire induction heating at a rate of approximately 15 °C/min. Before heat treatment, the air inside the furnace is evacuated to 10 Pa, then argon is backfilled to standard atmospheric pressure, and this process is repeated 2–3 times. During the heat treatment process, the gas pressure inside both furnaces is standard atmospheric pressure (approximately $1.01 \times 10^5$ Pa). Once the set temperature was reached, the specimens were held isothermally for a specified duration and then water-cooled to room temperature. The initial temperature of the cooling water is room temperature (approximately 20 °C). The cooling rate was estimated to be approximately 140 °C/s (it took approximately 6 s to cool from 860 °C to 20 °C). The setting of relevant heat treatment process parameters was based on the previous work [16].

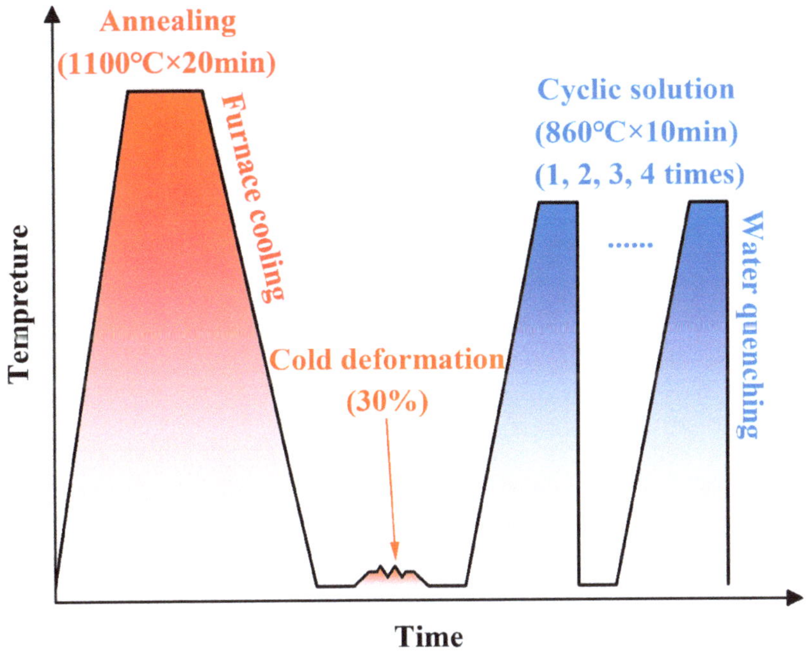


**Figure 1.** Process Flow Diagram.

The experimental steels subjected to 1–4 times of solution treatment after cold deformation were designated as specimens 1#–4#, respectively. Specimen 0# served as a comparative specimen, prepared via the traditional process (without cold deformation and with only one time solution treatment). Prior to solution treatment, the specimens for 0# and 1# were labeled as 0#-Before ST and 1#-Before ST, respectively.

### 2.2. Methods

Microstructural characterization and mechanical property testing were conducted on specimens in different states.

The microstructure of the materials was examined using a Zeiss Ultra Plus field emission scanning electron microscope (SEM) (Carl Zeiss AG, Oberkochen, Germany) equipped with an energy dispersive spectrometer (EDS) for elemental composition and content analysis. Specimens, sized at 10 mm × 10 mm × 5 mm, were ground, polished, etched with a 4% nitric acid-ethanol solution, and thoroughly cleaned with alcohol before SEM. During SEM analysis, an accelerating voltage of 20 kV was applied, and the back scattered electron mode (BSE) was chosen. Moreover, the secondary electron mode (SE) was employed to observe the morphology of the fracture surfaces of the tensile-tested specimens. Grain size was measured using Image J (Java 8) software.

An Empyrean X-ray diffractometer (XRD) was utilized to characterize the phases of the specimens. The specimen preparation for XRD was identical to that for SEM. The XRD system employed Cu K$\alpha$ radiation, with a 2$\theta$ scanning range of 30–90° and a scanning rate of 10°/min. Martensite and austenite were distinguished based on the diffraction peak angles corresponding to their respective crystallographic orientations, followed by statistical analysis of peak intensities. The XRD data were processed and analyzed using Origin 2022 software.

Room-temperature tensile properties of the specimens were evaluated using a Z050 electronic universal material testing machine (ZwickRoell, Ulm, Germany). The equipment had a maximum tensile load capacity of 50 kN and a tensile speed of 0.5 mm/min. As shown in Figure 2, the test specimens featured a parallel section with dimensions of 25 mm × 10 mm × 1 mm and a gauge length of 20 mm. Prior to testing, three 1 mm thick tensile test specimens were cut from a 3.5 mm thick plate. The surfaces and sides of the tensile specimens were ground to prevent stress concentration caused by cutting marks during the test. Three specimens were tested per experimental group to ensure the reliability of the mechanical property data.

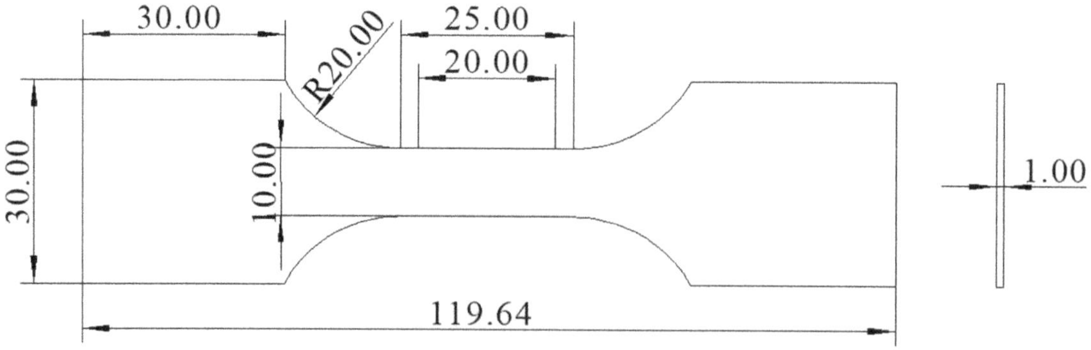


**Figure 2.** Schematic of the tensile specimen size (all measurements are in mm).

## 3. Results and Discussion

### *3.1. Microstructure and Composition Distribution*

#### 3.1.1. Microstructural Morphology

Figure 3 shows the microstructural morphologies of specimens 0#-Before ST and 1#-Before ST. Before solution treatment, the microstructure exhibits typical lath martensite with random orientation, and the original austenite grain boundaries are distinctly observable [17]. Martensite within the original austenite does not extend outward, with each martensite block parallel to one facet of the original austenite grain. Cold deformation is demonstrated to significantly modify the material's microstructural morphology, resulting in reduced sizes of both original austenite grains and martensite.

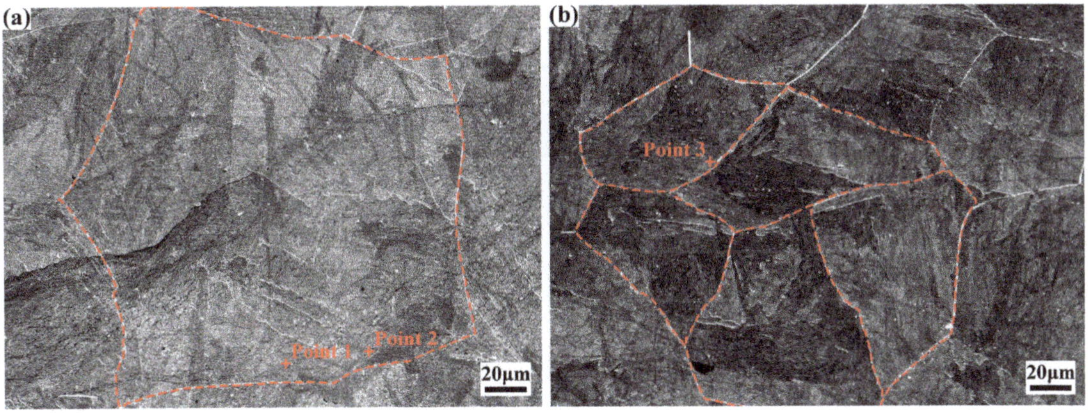


**Figure 3.** SEM images of the experimental steel before solution treatment: (**a**) 0#-Before ST; (**b**) 1#-Before ST.

Figure 3a illustrates specimen 0#-Before ST, which exhibits equiaxed original austenite grains with an axial length of approximately 188 μm. Figure 3b shows specimen 1#-Before ST subjected to 30% deformation, where multiple small austenite grains formed by the fragmentation of a single large grain are observed. These small grains display irregular morphologies, elongated along the rolling direction (73–125 μm in length) and correspondingly reduced in width (approximately 46–55 μm). Both specimens contain uniformly distributed tiny white particles with a diameter of 1–2 μm, with fewer particles present in 1#-Before ST. These particles are likely residual austenite formed during annealing.

The sizes of martensite and original austenite grains were quantitatively statistically analyzed using Image J software. Martensite length and width decreased from 76.1 μm to 53.9 μm and from 6.6 μm to 5.4 μm, respectively. The average area of original austenite grains reduced from 30,769 $\mu m^2$ to 8960 $\mu m^2$, with the corresponding equivalent circle radius decreasing from 99 μm to 53.4 μm.

Figure 4 illustrates the microstructural morphologies of specimens 0# and 1# of the experimental steel. After one time of solution treatment, the morphology of lath martensite is highly distinct with random orientation, while the original austenite grain boundaries exhibit blurring. Martensite dimensions were quantitatively analyzed using Image J software. Comparative analysis between specimens 0# and 1# revealed a reduction in martensite length from 47.0 μm to 18.3 μm and a corresponding decrease in martensite width from 4.2 μm to 2.8 μm. The average area of original austenite grains decreased significantly from 27,762 $\mu m^2$ to 317 $\mu m^2$, with the equivalent circle radius correspondingly reducing from 94 μm to 10 μm.

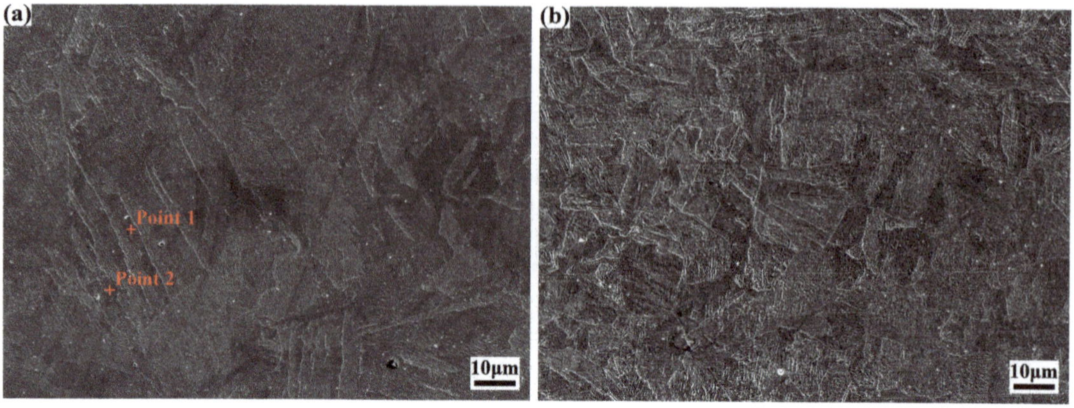


**Figure 4.** SEM images of the experimental steel after one cycle of solution treatment: (**a**) 0#; (**b**) 1#.

Notably, solution treatment further promoted grain refinement, with a more pronounced effect observed when solution treatment followed cold deformation. Additionally, significant changes in grain morphology were evident: most original austenite grains evolved into equiaxed morphology, which appears to suggest a recrystallization phenomenon. Trace amounts of fine white particles ($\approx$1 μm in size) were dispersedly distributed in both specimens, likely representing residual austenite formed during annealing and solution treatment.

Figure 5 displays the microstructural morphologies of specimens subjected to different times of solution treatments (Figure 5a is a partial magnification of Figure 4b). Distinct lath martensite structures and original austenite grains are observable. Notably, the martensite size decreases significantly with increasing solution treatment times, though a slight increase is evident after the fourth treatment. Image J software-based statistical analysis reveals that martensite length gradually decreases to 7.0 μm in specimen 3#, then increases slightly to 8.6 μm in specimen 4#, with corresponding width reduction to 1.3 μm. Specifically, martensite lengths in specimens 2#, 3#, and 4# are reduced by 26.8%, 61.7%, and 53.0% relative to specimen 1#, with widths decreasing by 46.4%, 53.6%, and 53.6%, respectively. Significantly, after four times of solution treatment, martensite length exhibits a minor increase rather than continuous reduction, while width remains unchanged.

In Figure 5a, the original austenite grains are relatively large, whereas those in Figure 5b show uneven sizes. By contrast, the original austenite grains in Figure 5c,d are smaller and more uniform. Image J software-based statistical analysis of original austenite grain sizes in the experimental steel subjected to 1, 2, 3, and 4 times of solution treatment revealed average grain areas of 317 $\mu m^2$, 70 $\mu m^2$, 34 $\mu m^2$, and 39 $\mu m^2$, with corresponding equivalent circle radii of 10.0 μm, 4.7 μm, 3.3 μm, and 3.5 μm, respectively. Compared to the specimen after 1 time of solution treatment, the martensite grain sizes of specimens after 2, 3, and 4 times of solution treatment were reduced by 53.0%, 67.0%, and 65.0%, respectively. It can be seen that with the increase in the number of solution treatments, the size of the original austenite grains first decreases and then slightly grows, which is consistent with the change rule of martensite size, indicating that grain refinement has an effect on refining the martensite structure. Meanwhile, some fine white granular structures with a size of about 1 μm can be observed, which may be retained austenite, and with the increase in the number of solution treatments, their volume fraction slightly increases.

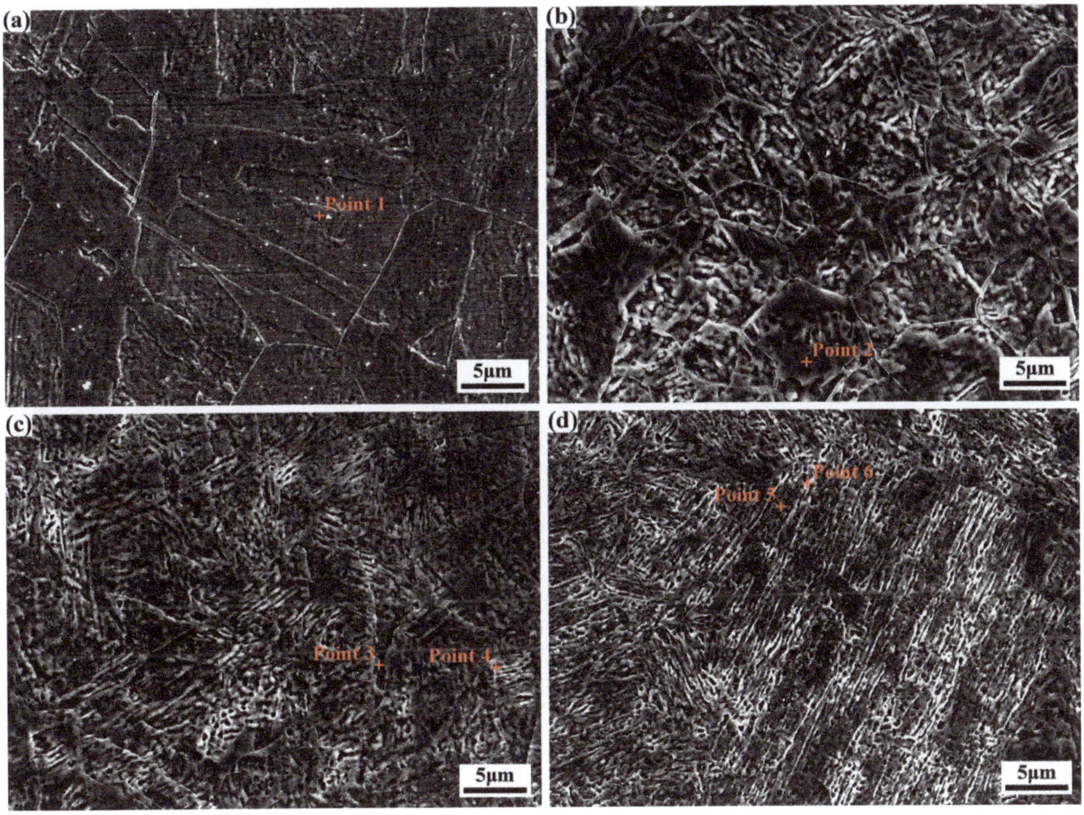


**Figure 5.** SEM images of the experimental steel after different times of solution treatment: (**a**) 1#; (**b**) 2#; (**c**) 3#; (**d**) 4#.

### 3.1.2. Microzone Composition

During the microscopic characterization described above, EDS point analysis was conducted on characteristic regions, with results tabulated in Table 2. For specimens before solution treatment and after one time of solution treatment, elemental distribution within the matrix remained relatively uniform, demonstrating that cold deformation and one time of solution treatment exerted negligible influence on the matrix elemental composition.

**Table 2.** EDS point analysis results of the experimental steel in different states.

| Specimen No. | EDS Point | Elemental Composition (wt.%) | | | | | |
|---|---|---|---|---|---|---|---|
| | | Ni | Co | Mo | Ti | Al | Fe |
| 0#-Before ST | Point 1 (Figure 3a) | 18.72 | 10.12 | 5.08 | 0.62 | 0.15 | Bal. |
| | Point 2 (Figure 3a) | 18.07 | 10.30 | 5.85 | 0.55 | 0.19 | Bal. |
| 1#-Before ST | Point 3 (Figure 3b) | 18.44 | 10.84 | 5.47 | 0.68 | 0.13 | Bal. |
| 0# | Point 1 (Figure 4a) | 18.81 | 10.41 | 5.77 | 0.68 | 0.16 | Bal. |
| | Point 2 (Figure 4a) | 18.57 | 10.03 | 5.86 | 0.75 | 0.17 | Bal. |
| 1# | Point 1 (Figure 5a) | 19.05 | 9.77 | 5.49 | 0.63 | 0.16 | Bal. |
| 2# | Point 2 (Figure 5b) | 18.93 | 9.56 | 5.34 | 0.66 | 0.16 | Bal. |
| 3# | Point 3 (Figure 5c) | 18.73 | 9.02 | 5.47 | 0.64 | 0.18 | Bal. |
| | Point 4 (Figure 5c) | 19.28 | 9.11 | 5.92 | 0.66 | 0.17 | Bal. |
| 4# | Point 5 (Figure 5d) | 18.74 | 8.94 | 5.66 | 0.63 | 0.16 | Bal. |
| | Point 6 (Figure 5d) | 19.53 | 9.07 | 5.93 | 0.69 | 0.19 | Bal. |

For specimens with varying times of solution treatments, the Ni and Co contents in the martensite matrix of specimens 1# and 2# were slightly higher than those in specimens 3# and 4#, though the overall elemental ratios remained similar. In specimens 3# and 4#, discrepancies in Ni and Mo contents were observed between the matrix and white granular structures. The Ni content in white granules (Points 4 and 6 in Figure 5) was marginally higher than that in the matrix (Points 3 and 5 in Figure 5), while other elements exhibited negligible content variations.

### 3.1.3. Phase Composition and Crystallographic Features

To further determine the white granular structures in the microstructure and quantify their phase composition, X-ray diffraction (XRD) analysis was conducted on experimental steel in various states. As illustrated in Figure 6, the diffraction data were processed using Jade 6 software, revealing that both 0# and 1# specimens consist exclusively of martensite ($\alpha'$) before and after solution treatment. The diffraction patterns are dominated by characteristic martensite peaks from the (110)$\alpha'$, (200)$\alpha'$, and (211)$\alpha'$ crystal planes, with their corresponding 2$\theta$ angles at approximately 44°, 65°, and 82°, respectively. Cold deformation induces a notable increase in the intensity of the (110)$\alpha'$ peak, accompanied by slight reductions in the (200)$\alpha'$ and (211)$\alpha'$ peak intensities. Significantly, the latter two peaks remain subdominant throughout, a trend consistently observed in the specimen after one time of solution treatment. Post-solution treatment, the (110)$\alpha'$ peak intensity further increases relative to the as-deformed state, while the (200)$\alpha'$ and (211)$\alpha'$ peaks continue to show marginal intensity decreases.

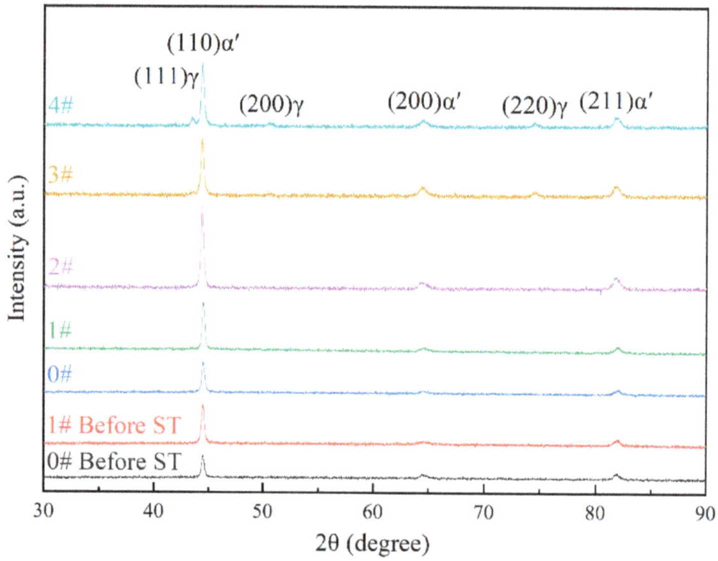


**Figure 6.** XRD patterns of experimental steel in different states.

Upon further comparison of the XRD phase analyses of specimens subjected to varying numbers of solution treatments, it is evident that both specimens 1# and 2# solely consist of the martensite phase. In contrast, when the number of solution treatments reaches 3 and 4, retained austenite emerges in the phase composition. A positive correlation is observed between the number of solution treatments and the intensity of the austenite diffraction peaks; specifically, as the number of solution treatments increases, the intensity of these peaks rises accordingly. In specimen 3#, the principal diffraction peak of austenite is identified as (220)γ, while the intensities of the (111)γ and (200)γ diffraction peaks are negligible. For specimen 4#, the dominant austenite diffraction peak shifts to (111)γ; the intensity of the (200)γ diffraction peak experiences a notable increase, whereas that of the (220)γ diffraction peak remains relatively stable. These findings strongly suggest that the white particles observed in the microstructure correspond to retained austenite. It should be noted that in certain specimens, the quantity of trace retained austenite is so minute (less than 1%) that it eludes detection during the XRD analysis.

To precisely quantify the relative fractions of the martensite and austenite phases, the diffraction peak intensities of each specimen were statistically analyzed via the Jade software. Subsequently, the relative phase fractions were calculated using the equations presented in Formulas (1) and (2) [18]:

$$V_{\alpha'} + V_{\gamma} = 1 \quad (1)$$

$$V_{\gamma} = \frac{1.4I_{\gamma}}{I_{\alpha'} + 1.4I_{\gamma}} \quad (2)$$

In the above formulas, $V_{\alpha'}$ and $V_{\gamma}$ are employed to denote the volume fractions of martensite and austenite, respectively. Correspondingly, $I_{\alpha'}$ and $I_{\gamma}$ signify the diffraction peak intensities of martensite and austenite. Specifically, the intensities of their principal diffraction peaks are considered, and in this context, the areas of the diffraction peaks are utilized for analysis. For specimen 3#, the principal diffraction peaks are identified as (110)α′ and (220)γ. In contrast, for specimen 4#, the principal diffraction peaks are (110)α′ and (111)γ.

The outcomes of the analysis are presented in Table 3. No austenite diffraction peaks were discerned in specimens 1# and 2#, indicating a relative austenite content of 0%. Specimen 3# exhibited an austenite relative content of 4.5%, while specimen 4# had an austenite relative content of 8.9%.

**Table 3.** Relative phase contents of the tested steel after different times of solution treatment.

| Specimen Number | Intensity of Principal Diffraction Peaks | | Phase Content (%) | |
|---|---|---|---|---|
| | (110)α′ | (111)γ/(220)γ | Martensite | Austenite |
| 1# | 28,088 | 0 | 100 | 0 |
| 2# | 53,976 | 0 | 100 | 0 |
| 3# | 25,866 | 874 | 95.5 | 4.5 |
| 4# | 30,797 | 2144 | 91.1 | 8.9 |

In consideration of the potential occurrence of texture induced by cold deformation, the test result data were subjected to crystal plane texture analysis. The relative texture coefficients of crystal planes corresponding to each experimental steel were calculated by adopting Formula (3) [19]:

$$RTc(hkl) = \frac{I_{(hkl)} / I_{0_{(hkl)}}}{\sum_{i=0}^{n} I_{(hkl)} / I_{0(hkl)}} \quad (3)$$

Within the formula, $RTc(hkl)$ denotes the relative texture coefficient of the crystal plane. Specifically, $I_{(hkl)}$ represents the experimentally measured diffraction peak intensity of the specimen, while $I_{0_{(hkl)}}$ corresponds to the standard diffraction peak intensity documented in the JCPDS card. Additionally, $n$ stands for the quantity of diffraction peaks.

Table 4 showcases the computational outcomes of the relative texture coefficients of crystal planes for the experimental steel across diverse states. Evidently, in all specimens, the crystal planes with the (110)α′ orientation hold the highest proportion. Cold deformation leads to a significant increase in the relative texture coefficient of the (110)$\alpha'$ crystal plane, which rises from 37.5% in specimen 0# before aging treatment (0#-Before ST) to 52.0% in specimen 1# before aging treatment (1#-Before ST). Following one time of solution treatment, the relative texture coefficients of the (110)$\alpha'$ crystal plane for specimens 0# and 1# ascend to 46.6% and 55.6%, respectively. This phenomenon will trigger a preferred orientation of the material along the (110)$\alpha'$ crystal plane, consequently influencing the material's practical application.

**Table 4.** Relative texture coefficients of crystal planes of the experimental steel under different conditions (%).

| Specimen Number | *RTc(hkl)* | | |
|---|---|---|---|
| | **(110)$\alpha'$** | **(200)$\alpha'$** | **(211)$\alpha'$** |
| 0#-Before ST | 37.5 | 29.8 | 32.7 |
| 1#-Before ST | 52.0 | 20.1 | 27.9 |
| 0# | 46.6 | 26.7 | 26.7 |
| 1# | 55.6 | 20.1 | 24.3 |
| 2# | 50.2 | 24.6 | 25.2 |
| 3# | 40.8 | 32.4 | 26.8 |
| 4# | 45.9 | 28.5 | 25.6 |

Nonetheless, as the number of solution treatments rises, the relative texture coefficient of the (110)$\alpha'$—oriented crystal plane exhibits a declining tendency. Conversely, the relative texture coefficients of the (200)$\alpha'$—and (211)$\alpha'$—oriented crystal planes witness a corresponding increase. After three times of solution treatments, the relative texture coefficients of crystal planes in each orientation approach consistency, and the texture phenomenon gradually fades away.

### *3.2. Mechanism of Grain Refinement and Equiaxialization*

Microstructural characterization (Figures 2–4) demonstrates that cold deformation coupled with cyclic solution treatment effectively refines the original austenite grains and martensitic structure (statistical data are tabulated in Table 5). The optimal grain refinement is attained after three times of solution treatment subsequent to cold deformation, during which the grains gradually evolve toward an equiaxed morphology.

**Table 5.** Microstructural dimensions of the experimental steel under different conditions.

| Specimen Number | Lath Martensite | | Original Austenite Grain | |
|---|---|---|---|---|
| | Length (μm) | Width (μm) | Average Area (μm$^2$) | Equivalent Circle Radius (μm) |
| 0# | 47 | 4.2 | 27,762 | 94 |
| 1# | 18.3 | 2.8 | 317 | 10 |
| 2# | 13.4 | 1.5 | 70 | 4.7 |
| 3# | 7 | 1.3 | 34 | 3.3 |
| 4# | 8.6 | 1.3 | 39 | 3.5 |

First, cold deformation induced fragmentation of the original austenite grains in the annealed experimental steel, accompanied by synchronous refinement of the internal martensite dimensions. Upon application of a specific deformation magnitude to the original austenite grains, a series of slip bands and high-density dislocations formed within the grains, progressively subdividing the original austenite into subgrains. With cumulative deformation, subgrain boundaries underwent gradual transformation into high-angle grain boundaries, thereby facilitating the formation of new grains and achieving mechanical fragmentation-induced refinement of the original austenite structure [10].

Based on the XRD characterization results, the Williamson-Hall method (i.e., Formulas (4) and (5)) was employed to quantify the dislocation density of the experimental steel under different processing states [20]:

$$\beta\cos\theta = \frac{K\lambda}{D} + 4\varepsilon\sin\theta \tag{4}$$

$$\rho = \frac{14.4\varepsilon^2}{b^2} \tag{5}$$

In the formulas, $\theta$ denotes the angle corresponding to each diffraction peak, which must be converted to radians during calculations; $\beta$ represents the full width at half maximum (FWHW) of each diffraction peak, similarly converted to radians for computation; $K$ is the Scherrer constant, taking a value of 0.9 when $\beta$ corresponds to the FWHW; $\lambda$ is the X-ray wavelength, with Cu $K\alpha_1$ set as 0.15406 nm; $D$ signifies the grain size; $\varepsilon$ is the microstrain; $\rho$ represents the dislocation density; and $b$ is the Burgers vector, specified as 0.249 nm for the martensite matrix. The $\theta$ and $\beta$ values for each diffraction peak were derived by fitting the corresponding XRD data using Origin software. Substituting the $\theta$ and $\beta$ values associated with the three diffraction peaks $(110)\alpha'$, $(200)\alpha'$, and $(211)\alpha'$ of the specimen into Formula (4) generated a system of binary linear equations, from which the values of $D$ and $\varepsilon$ were solved. These parameters were subsequently substituted into Formula (5) to compute the $\rho$ value. The dislocation density calculation results were tabulated in Table 6.

**Table 6.** Dislocation Density Calculation Results.

| Specimen Number | (hkl) | $\theta$ (°) | $\beta$ (°) | $\varepsilon$ | $\rho$ ($nm^{-2}$) |
|---|---|---|---|---|---|
| 0#-Before ST | $(110)\alpha'$ | 44.50607 | 0.34703 | $2.89 \times 10^{-3}$ | $1.96 \times 10^{-3}$ |
| | $(200)\alpha'$ | 64.65564 | 0.97128 | | |
| | $(211)\alpha'$ | 81.95592 | 0.62215 | | |
| 1#-Before ST | $(110)\alpha'$ | 44.48197 | 0.35606 | $4.14 \times 10^{-3}$ | $4.01 \times 10^{-3}$ |
| | $(200)\alpha'$ | 64.71126 | 1.66797 | | |
| | $(211)\alpha'$ | 81.95428 | 0.67275 | | |
| 0# | $(110)\alpha'$ | 44.49842 | 0.36194 | $3.16 \times 10^{-3}$ | $2.34 \times 10^{-3}$ |
| | $(200)\alpha'$ | 64.60146 | 0.9415 | | |
| | $(211)\alpha'$ | 81.98378 | 0.67046 | | |
| 1# | $(110)\alpha'$ | 44.50395 | 0.33175 | $1.94 \times 10^{-3}$ | $0.88 \times 10^{-3}$ |
| | $(200)\alpha'$ | 64.62762 | 0.84887 | | |
| | $(211)\alpha'$ | 81.9837 | 0.53127 | | |

The results reveal that prior to solution treatment, cold deformation enhanced microstrain and dislocation density. Specifically, the dislocation density increased from $1.96 \times 10^{-3}$ $nm^{-2}$ in the specimen 0#-Before ST to $4.01 \times 10^{-3}$ $nm^{-2}$ in the specimen 1#-Before ST, representing a 104.6% increase. For the specimen 0# without cold deformation, the dislocation density after solution treatment ($2.34 \times 10^{-3}$ $nm^{-2}$) exhibited a marginal increase compared to its pre-solution treatment state. This phenomenon arises because the absence of cold deformation resulted in insufficient dislocation density during the austenitization heating stage of the solution treatment process to satisfy recrystallization criteria. Consequently, during the cooling stage of the solution treatment process, the shear-type diffusionless phase transformation induced an elevation in dislocation density. In contrast, the dislocation density of specimen 1# decreased to below $1 \times 10^{-3}$ $nm^{-2}$ after solution treatment. Coupled with grain size statistics derived from microstructural morphology, it can be inferred that recrystallization occurred in the cold-deformed specimen 1# in the process of solution treatment.

During the heating stage of the solution treatment process, the austenite grain size is governed by two competitive processes: grain nucleation and growth. Cold deformation introduces an increase in dislocation density and high-angle grain boundaries, accompanied by irregular grain boundary morphology, which collectively provide abundant nucleation sites for austenitization. With prolonged heating, grain growth proceeds gradually. With high nucleation density facilitating rapid stabilization of grain boundary migration, at the austenitization start temperature (800 °C), complete transformation to austenitic structure is achieved. Upon further heating to the recrystallization temperature regime (800–900 °C), the inherited high-density dislocations and deformation-stored energy within grains create favorable thermodynamics for recrystallization nucleation, yielding new fine non-deformed grains that ultimately evolve into a uniform equiaxed microstructure [11].

In conclusion, cold deformation induces high-density dislocations and deformation-stored energy, which synergistically enhance austenitization nucleation efficiency and stabilization kinetics during the first solution treatment. Concurrently, these factors promote recrystallization, culminating in grain refinement and equiaxialization.

Secondly, cyclic solution treatment is capable of further refining the original austenite grains, consequently leading to the refinement of the martensite structure. During the water-cooling, a phase transformation from austenite to martensite takes place. The volumetric disparity between the two phases gives rise to the generation of dislocations. These dislocations are subsequently inherited in the subsequent solution treatment. When combined with the defects and deformation-stored energy, which were generated during the cold deformation process and remained uneliminated in the previous solution treatment, they continue to serve as the driving force for austenitization nucleation and recrystallization [21]. Consequently, the grains undergo further refinement.

Nevertheless, the grain-refining effect gradually diminishes with an increasing number of solution treatments. Evidently, this is because the quantity of dislocations generated by the phase transformation is significantly lower than that produced by cold deformation, which also highlights the significance of cold deformation in grain and structure refinement. The optimal grain refinement is attained after the third solution treatment. If the solution treatment is continued thereafter, the grains will not be refined; rather, they will experience a slight growth. This phenomenon primarily stems from the exhaustion of the defects and deformation-stored energy generated by cold deformation after multiple solution treatments, and the dislocations solely generated by the phase transformation are inadequate to supply the driving force necessary for austenitization nucleation and recrystallization.

### 3.3. Generation and Elimination of Grain Preferred Orientation

The relative texture coefficients of the crystal planes of the experimental steel under different states, as presented in Table 4, indicate that cold deformation induces the preferred orientation of crystals. This preferred orientation can be effectively eliminated through subsequent cyclic solution treatment. During cyclic solution treatment, re-austenitization and recrystallization take place, resulting in the consumption of micro-defects generated by cold deformation. In this process, new, non-deformed grains nucleate by consuming the deformation-stored energy. The equiaxed growth mechanism of these grains effectively eradicates the original crystallographic preferred orientation characteristics. As a fully recrystallized structure forms, the material ultimately exhibits an equiaxed grain structure with isotropic distribution, and the original deformation texture is entirely eliminated.

The XRD data corroborate the aforementioned view. As the number of solution treatments increases, the relative texture coefficient of the $(110)\alpha'$ crystal plane of the tested steel diminishes, whereas the relative texture coefficients of the $(200)\alpha'$ and $(211)\alpha'$ crystal planes increase. The relative texture coefficients of these three planes gradually converge, which signifies the progressive completion of recrystallization and also denotes the gradual reduction and eventual disappearance of the texture within the structure.

The existence of cold deformation texture confers certain advantages to grain refinement during cyclic solution treatment [22–25]. During austenitization, the distinct atomic arrangements and inter-atomic spacings among different crystal plane families, along with the crystal orientation disparities induced by the texture, influence the dislocation density and distribution. This situation is favorable for austenite nucleation, with these regions acting as preferential nucleation sites. Furthermore, in materials exhibiting texture, the orientation constrains the migration of grain boundaries, impeding their movement and thereby suppressing grain growth.

The correlation between the data presented in Tables 4 and 5 further validates this perspective. After undergoing cold deformation and three solution treatments, the 3# experimental steel exhibits the least pronounced texture. Concurrently, its grains and microstructure attain the highest level of refinement, which further attests to the rationality of the cold deformation—cyclic solution treatment process devised in this study.

### *3.4. Formation Mechanism of Retained Austenite*

Cold deformation treatment exerts a non-negligible influence on the phase composition. XRD analysis reveals that the experimental steel before solution treatment consists exclusively of martensite phase; however, SEM observation indicates the presence of trace amounts of fine, white, dot-like retained austenite in the microstructure. Notably, the cold-deformed specimen 1#-Before ST exhibits a significantly lower density of these white dot-like features. As documented in Ref. [26], cold deformation triggers the transformation of retained austenite into martensite, a phenomenon associated with the transformation-induced plasticity (TRIP) effect. The underlying mechanism can be attributed to the inherently low stability of retained austenite, which requires a critical phase transformation driving force to convert into martensite. Cold deformation provides the necessary mechanical driving force to surmount this energy barrier, thereby promoting the transformation from retained austenite to martensite and consequently reducing the retained austenite content.

Upon multiple solution treatments of cold-deformed experimental steel, its phase composition transitions from a monophasic martensite structure to a biphasic mixture of martensite and retained austenite. Significantly, Table 3 demonstrates a positive correlation between the number of solution treatments and retained austenite content, with higher numbers of solution treatments corresponding to increased retained austenite retention.

The emergence of retained austenite is attributed to local Ni partitioning during cyclic solution treatments [27]. Diffusion processes, governed by chemical potential gradients, inherently proceed in the direction of decreasing chemical potential, where chemical potential is a function of both elemental concentration and temperature. Owing to the lower chemical potential of Ni in the austenite phase relative to martensite, Ni atoms migrate from martensitic domains to adjacent austenitic regions, creating Ni-enriched austenite and Ni-depleted martensite zones. This mechanism is corroborated by EDS point analysis (Table 2), which evidences slightly elevated Ni concentrations in retained austenite compared to martensite.

As a potent austenite-stabilizing element, Ni reduces the Ms temperature in a concentration-dependent manner. When local Ni enrichment surpasses a certain amount, the Ms temperature is depressed below room temperature, preventing transformation of Ni-rich austenite to martensite during water quenching. Consequently, these regions persist as retained austenite. This mechanistic framework explains both the appearance of retained austenite in the microstructure of maraging steel after cyclic solution treatments and the monotonic increase in retained austenite content with the number of solution treatments.

## *3.5. Mechanical Properties and Strengthening-Plasticizing Mechanism*

### 3.5.1. Mechanical Behavior

The microstructural evolution during aging of maraging steel is primarily characterized by the precipitation of alloy compounds and the formation of reverse transformation austenite, whereas the grain morphology and martensitic structure in the as-solutionized state remain essentially unaltered [28–30]. This highlights the critical need to investigate the mechanical properties and strengthening-plasticizing mechanisms of the solution-treated experimental steel for evaluating the final aged-state performance, with particular emphasis on plasticity. High-level plasticity in the solutionized condition serves as a foundational prerequisite for enhancing the overall ductility of maraging steel.

Room-temperature tensile testing was performed on the solution-treated specimens. Figure 7 illustrates the true stress-strain curves for each specimen, with corresponding mechanical property parameters tabulated in Table 7. Notably, cold deformation followed by solution treatment induced a significant enhancement in both strength and elongation. With increasing numbers of solution treatments, strength exhibited a non-monotonic trend, first increasing and then decreasing. The yield strength peaked at 881 MPa after two times of solution treatments, while the ultimate tensile strength reached a maximum of 1240 MPa after three times of solution treatments. Concurrently, elongation followed a similar trend, achieving a peak value of 9.93% after three times of solution treatments—nearly double the value of specimen 0#. Comprehensive performance evaluation indicates that the experimental steel subjected to cold deformation and three times of solution treatments displayed optimal room-temperature tensile properties.

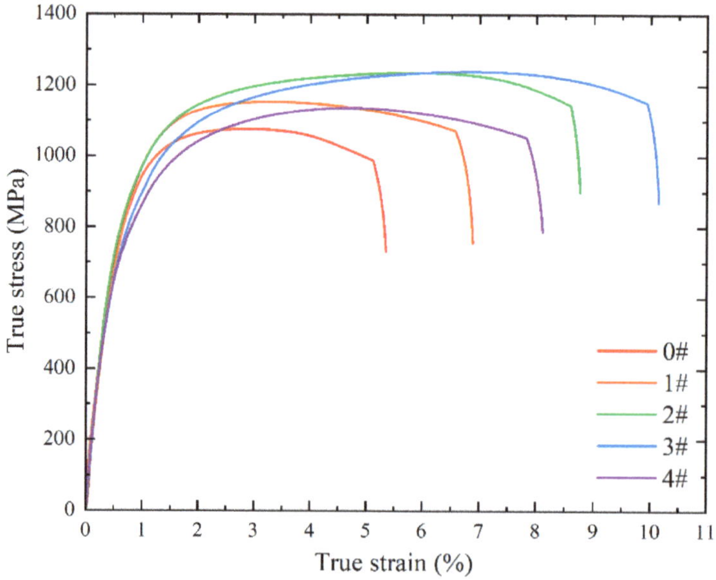


**Figure 7.** Room-temperature tensile true stress-strain curves of experimental steel in different states.

**Table 7.** Room-temperature tensile mechanical property parameters of experimental steel in different states.

| Specimen Number | Yield Strength (MPa) | Ultimate Tensile Strength (MPa) | Elongation (%) |
|---|---|---|---|
| 0# | 853 | 1076 | 5.11 |
| 1# | 873 | 1153 | 6.57 |
| 2# | 881 | 1236 | 8.59 |
| 3# | 813 | 1240 | 9.93 |
| 4# | 778 | 1136 | 7.82 |

#### 3.5.2. Strengthening Mechanism

In subsequent aging treatments, although changes in solid solution content, precipitated phases, and phase composition are induced, the original austenite grain structure and retained austenite phase remain fundamentally unaltered. Therefore, this paper focuses exclusively on the effects of these two microstructural features—original austenite grain size and retained austenite content—on the mechanical strength, with theoretical support from the fine-grain strengthening model (Formula (6)) [31]:

$$\sigma_{gb} = \sigma_0 + k_y d^{-1/2} \quad (6)$$

In the formula, $\sigma_0$ denotes the matrix yield strength, with a typical value of 50 MPa for iron-based materials; $k_y$ presents the slope of the $\sigma_y - d^{-1/2}$ re relationship, specified as 8.8 MPa·mm$^{1/2}$ for 18Ni maraging steel [32]; and $d$ corresponds to the average grain diameter. Converting martensite size to equivalent circular diameter for ease of calculation. The calculated fine-grain strengthening contributions for the experimental steels are 120 MPa, 148 MPa, 174 MPa, 201 MPa, and 193 MPa, respectively, as shown in Table 8. When compared with the corresponding tensile test strength data, these contributions exhibit a strong correlation with the strength increments. Cold deformation and cyclic solution treatment primarily affect the experimental steel by inducing grain refinement. Microstructural and mechanical property analyses reveal that as the number of solution treatments increases, the grain size first decreases substantially and then undergoes a slight increase, while the room-temperature tensile strength follows a corresponding trend of initial increase followed by decrease.

**Table 8.** The fine-grain strengthening contributions of experimental steel in different states.

| Specimen Number | Average Grain Diameter (μm) | The Fine-Grain Strengthening Contribution (MPa) |
|---|---|---|
| 0# | 15.86 | 120 |
| 1# | 8.08 | 148 |
| 2# | 5.06 | 174 |
| 3# | 3.4 | 201 |
| 4# | 3.77 | 193 |

The strength of retained austenite is significantly lower than that of martensite, a characteristic that inherently tends to diminish the overall strength of the material. Notably, however, when compared to the 2# single-phase martensite specimen, the ultimate tensile strength of the specimen 3# did not decline with the introduction of austenite; instead, it reached a peak value of 1240 MPa. This phenomenon is attributed to the substantial strength contribution from grain refinement, which sufficiently counterbalanced the strength-degrading effect of the minor retained austenite phase. In the case of specimen 4#, where the retained austenite content reached 8.9%, the ultimate tensile strength decreased to 1136 MPa. Despite possessing finer grains and a more refined microstructure compared to the 1# and 2# specimens, the strength enhancement from grain refinement in the specimen 4# only marginally offset the detrimental impact of retained austenite. These results underscore that minimizing the content of retained austenite is critical for optimizing the strength of maraging steel, as the beneficial effects of grain refinement become insufficient to compensate for the strength loss once retained austenite exceeds a critical threshold.

#### 3.5.3. Plasticization Mechanism

The SEM fractographies of the tensile fracture surfaces for the experimental steel after solution treatment are presented in Figure 8. Specifically, Figure 8a illustrates a mixed fracture mode characterized by dominant ductile fracture with localized brittle fracture features. The surface displays dimples of non-uniform sizes and depths, with white tearing ridges distributed adjacent to both dimples and cleavage planes. The presence of dimples signifies that the specimen underwent substantial plastic deformation prior to fracture, whereby energy is dissipated through plastic deformation, thereby suppressing crack propagation.

Figure 8b exemplifies typical ductile fracture behavior, where dimples occupy nearly the entire fracture surface, and numerous submicron-scale microvoids are observed at the base of the dimples. This morphology indicates that fracture initiated and propagated via microvoid coalescence and growth mechanisms. With an increase in the number of solution treatments, all specimens' tensile fracture surfaces predominantly exhibit ductile fracture characteristics. As shown in Figure 8c,d, the dimples are highly uniform, with the latter showing smaller dimple dimensions. Conversely, Figure 8e reveals a less regular fracture surface featuring larger zigzag voids and distinct brittle fracture zones, indicative of a mixed ductile-brittle fracture mode.

A comprehensive comparison reveals a remarkable correlation between the fracture morphology and the elongation of the specimens. For the specimen 0#, it features a relatively large proportion of dimple area and a small proportion of cleavage plane area. This distinct morphological characteristic is in excellent agreement with the 5.11% elongation obtained from the room-temperature tensile test. The specimen 1# demonstrates a notable plastic recovery phenomenon. As the proportion of dimples significantly increases and the cleavage planes are nearly absent, the specimen achieves an elongation of 6.57%. When it comes to the specimens that have undergone 2 and 3 times of solution treatment, their fracture morphologies are solely composed of dimples. In particular, the dimples in the specimen 3# treated with 3 times the solution treatment are more uniformly distributed compared to those in the specimen 2#. This more uniform dimple distribution is responsible for the optimal plasticity, with an elongation reaching 9.93%.

The microstructural features dictate both the fracture morphology and the plastic behavior of the material. First, grain refinement enhances the uniformity of plastic deformation and intergranular deformation coordination. Finer grains increase the grain boundary area, thereby elevating the probability of dislocation pile-up at boundaries. This mechanism promotes coordinated deformation between adjacent grains and homogenizes dislocation distribution within each grain, enabling more uniform plastic deformation throughout the material matrix. Second, uniformly equiaxed grain structures exhibit isotropic dimensional homogeneity, leading to balanced resistance against dislocation motion in all crystallographic directions during plastic deformation. This isotropy facilitates dislocation slip and multiplication, allowing the material to deform uniformly across all axes and mitigating anisotropy-induced local stress concentrations. Third, texture elimination removes restrictions on dislocation movement along specific crystallographic planes/directions, further enhancing deformation coordination and overall plasticity.

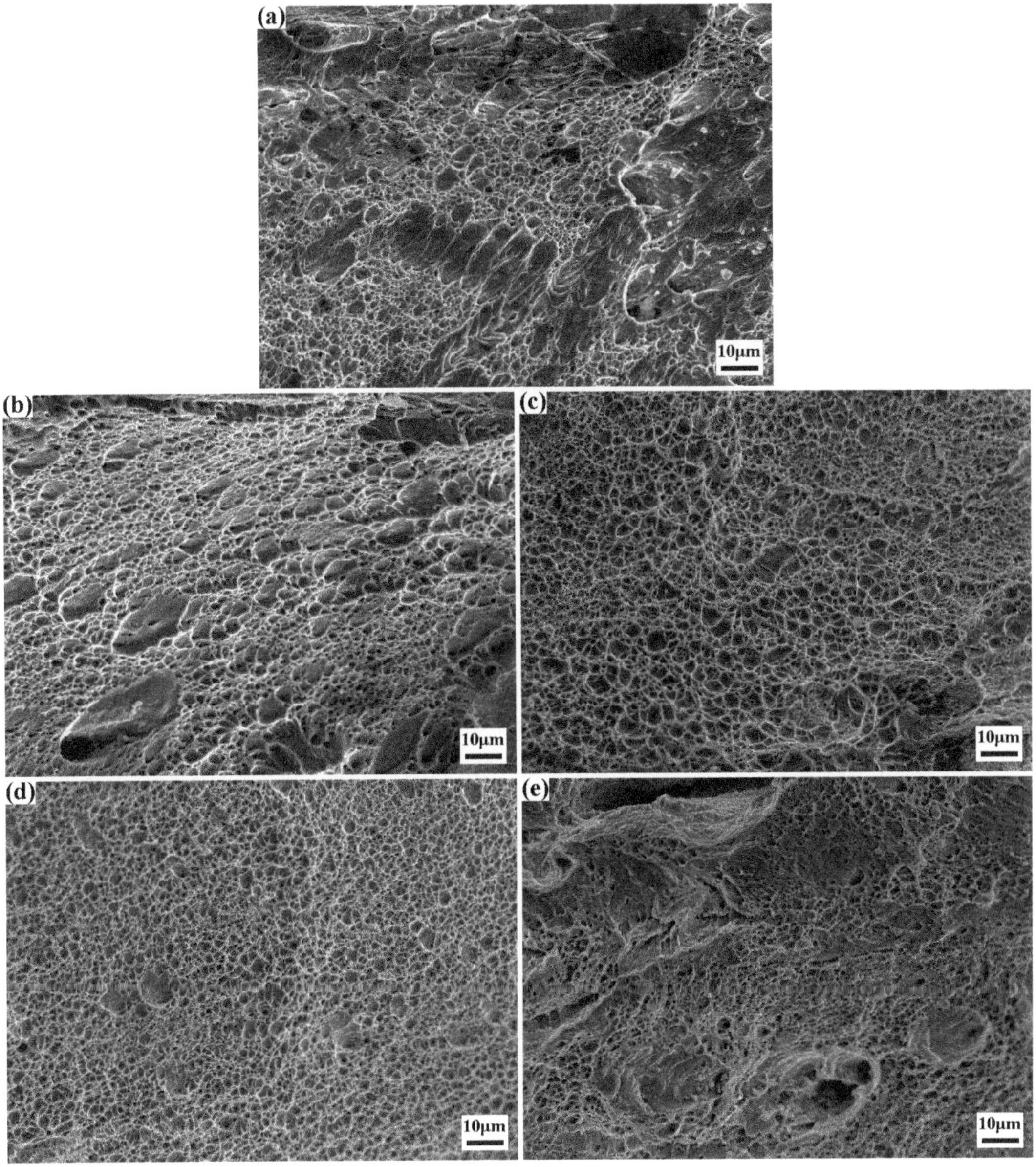


**Figure 8.** SEM morphologies of tensile fracture surfaces of experimental steel after solution treatment: (**a**) 0#; (**b**) 1#; (**c**) 2#; (**d**) 3#; (**e**) 4#.

Notably, retained austenite—owing to its ductile nature—typically contributes positively to plasticity. In specimen 3#, the combined effects of 4.5% retained austenite and grain refinement synergistically maximize plasticity. However, in the specimen 4# subjected to four times of solution treatment, despite an 8.9% retained austenite content, grain coarsening dominates, causing elongation to decline. This underscores that the plastic benefit of minor retained austenite is subordinate to grain refinement, with the latter serving as the primary determinant of ductility.

## 4. Conclusions

Through SEM and XRD microstructural characterization and room-temperature tensile testing, the influence of cold deformation-cyclic solution treatment on the microstructure and mechanical properties of 18Ni(300) maraging steel prior to aging was systematically investigated. The key findings are summarized as follows:

(1) Compared with the traditional process, the cold deformation-cyclic solution treatment process can effectively refine the original austenite grains (minimum equivalent circle radius: 3.3 μm) and martensitic microstructure (minimum length and width: 7 μm and 1.3 μm, respectively) and render the original austenite grains uniformly equiaxed.

(2) Cold deformation introduces crystallographic preferred orientation (texture), which is mitigated through recrystallization during cyclic solution treatments. The interplay between texture generation and annihilation enhances grain refinement efficiency.

(3) Cyclic solution treatment promotes the formation of retained austenite within the martensitic matrix, with its volume fraction increasing monotonically with treatment cycles. After three cycles, the retained austenite content was 4.5%, and after four cycles, the retained austenite content was 8.9%.

(4) Grain refinement and uniform equiaxialization are the dominant strengthening-toughening mechanisms, contributing to concurrent improvements in strength and ductility. In contrast, retained austenite acts as a soft phase, reducing matrix strength while providing limited plasticity enhancement, necessitating strict compositional control.

(5) Under fixed solution parameters, the optimal number of solution treatments is deformation-dependent. For 30% cold deformation, three solution cycles at 860 °C for 10 min yield the most refined microstructure, fully equiaxed grains, complete texture elimination, and balanced retained austenite. This condition achieves the ideal strength-ductility balance (tensile strength of 1240 MPa and elongation of 9.93%), establishing a robust microstructural foundation for subsequent aging treatment.

**Author Contributions:** Conceptualization: F.H. and D.L.; methodology, Z.H.; software, W.Z. and Z.C.; validation, Z.C.; formal analysis, F.H.; investigation, F.H. All authors have read and agreed to the published version of the manuscript.

**Funding:** This research was funded by [National Natural Science Foundation of China] grant number [52374399], [Open Fund of the Hubei Longzhong Laboratory] grant number [2024KF-15], and [High-level talents start-up project, West Anhui University] grant number [WGKQ2022016].

**Data Availability Statement:** The original contributions presented in this study are included in the article. Further inquiries can be directed to the corresponding authors.

**Conflicts of Interest:** Author Feng Huang was employed by the company Hubei Longzhong Laboratory. The remaining authors declare that the research was conducted in the absence of any commercial or financial relationships that could be construed as a potential conflict of interest.

## References

1. Wan, J.; Ruan, H.; Ding, Z.; Kong, L.B. A novel maraging stainless steel ultra-high-strengthened by multi-nanoprecipitations. *Scr. Mater.* **2023**, *226*, 115224. [CrossRef]
2. Yang, M.; Zhang, D.; Yao, Z.; Ma, Z.; Luan, J.; Wang, C.; Kuhn, B.; Jiao, Z.; Zhao, Y.; Yang, T.; et al. Precipitation behavior of the G-phase strengthened 7Ni maraging steels. *J. Mater. Res. Technol.* **2023**, *26*, 9261–9275. [CrossRef]
3. Li, J.; Zhan, D.; Jiang, Z.; Zhang, H.; Yang, Y.; Zhang, Y. Progress on improving strength-toughness of ultra-high strength martensitic steels for aerospace applications: A review. *J. Mater. Res. Technol.* **2023**, *23*, 172–190. [CrossRef]
4. Niu, M.C.; Chen, C.J.; Li, W.; Yang, K.; Luan, J.H.; Wang, W.; Jiao, Z.B. Atomic-scale understanding of solute interaction effects on grain boundary segregation, precipitation, and fracture of ultrahigh-strength maraging steels. *Acta Mater.* **2023**, *253*, 118972. [CrossRef]

5. Mooraj, S.; Feng, S.; Luebbe, M.; Register, M.; Liu, J.; Li, T.; Yavas, B.; Schmidt, D.P.; Priddy, M.W.; Nicholas, M.B.; et al. Martensitic transformation induced strength-ductility synergy in additively manufactured maraging 250 steel by thermal history engineering. *J. Mater. Sci. Technol.* **2025**, *211*, 212–225. [CrossRef]
6. Yiran, W.; Ruian, W.; Yimin, G. Effect of Co content on microstructure and mechanical properties of maraging steel. *J. Mater. Res. Technol.* **2023**, *27*, 3887–3899. [CrossRef]
7. Chen, T.; Chang, R.; Cheng, Z.; Qian, B.; Qiu, J.; Liang, Z.; Ning, X.; Sun, S.; Liu, B. Selective laser melting fabricated carbon nanotube reinforced 18Ni300 maraging steel directly used for mold repairing. *Vacuum* **2024**, *228*, 113448. [CrossRef]
8. Kasana, S.S.; Pandey, O.P. Effect of multipass cold plastic deformation and heat treatment on the mechanical properties of C250 marageing steel for aerospace applications. *Proc. Inst. Mech. Eng. Part L J. Mater. Des. Appl.* **2021**, *235*, 1551–1566. [CrossRef]
9. Lian, Y.; Huang, J.; Zhang, J.; Zhao, C.; Gao, W.; Zhang, Z.; Ma, M. Effects of cold rolling on the microstructure and properties of Fe-Cr-Ni-Mo-Ti maraging steel. *Mater. Sci. Eng. A* **2018**, *712*, 663–670. [CrossRef]
10. Huang, J.; Xu, Z. Evolution mechanism of grain refinement based on dynamic recrystallization in multiaxially forged austenite. *Mater. Lett.* **2006**, *60*, 1854–1858. [CrossRef]
11. Rosenauer, A.; Brandl, D.; Ressel, G.; Hönigmann, T.; Stadler, M.; Turk, C.; Gammer, C.; Wiessner, M.; Gamsjäger, E.; Stockinger, M.; et al. Multiscale in-situ observations of the micro- and nanostructure of a PH 13-8 Mo maraging steel during austenitization. *Mater. Charact.* **2024**, *208*, 113554. [CrossRef]
12. Xiao, K.; Han, S.; Li, Z.; Geng, R.; Han, G.; Li, Y.; Wang, C. Influence of Cyclic Heat Treatment Temperature on Microstructure and Mechanical Properties of 18Ni(C250) Maraging Steel. *Materials* **2024**, *17*, 2796. [CrossRef]
13. Li, Y.; Yu, J.; Li, S.; Wang, S.; Ren, Y.; Yang, K.; Wang, Y.-D. The influence of post-aging treatment on the microstructure and micromechanical behaviors of additively manufactured maraging steel investigated by in situ high-energy X-ray diffraction. *J. Mater. Sci. Technol.* **2024**, *200*, 1–12. [CrossRef]
14. Tariq, F.; Shifa, M.; Baloch, R.A. Effect of Overaging Conditions on Microstructure and Mechanical Properties of Maraging Steel. *Met. Sci. Heat Treat.* **2020**, *62*, 188–194. [CrossRef]
15. Feitosa, A.L.M.; Escobar, J.; Ribamar, G.G.; Avila, J.A.; Padilha, A.F. Direct Observation of Austenite Reversion During Aging of 18Ni (350 Grade) Maraging Steel Through In-Situ Synchrotron X-Ray Diffraction. *Met. Mater. Trans. A* **2021**, *53*, 420–431. [CrossRef]
16. Huang, F.; Cheng, Z.; Zhang, D.; Qian, D.; Liu, Y.; Hu, Z.; Hua, L. A novel process for simultaneously improving the strength and plasticity of 18Ni(350) maraging steel. *J. Mater. Res. Technol.* **2024**, *33*, 6144–6156. [CrossRef]
17. Viswanathan, U.; Dey, G.; Sethumadhavan, V. Effects of austenite reversion during overageing on the mechanical properties of 18 Ni (350) maraging steel. *Mater. Sci. Eng. A* **2005**, *398*, 367–372. [CrossRef]
18. Fan, Y.H.; Zhang, B.; Yi, H.L.; Hao, G.S.; Sun, Y.Y.; Wang, J.Q.; Han, E.-H.; Ke, W. The role of reversed austenite in hydrogen embrittlement fracture of S41500 martensitic stainless steel. *Acta Mater.* **2017**, *139*, 188–195. [CrossRef]
19. Liang, X.; Chen, X.; Zhai, Z.; Huang, R.; Yu, T.; Yin, S. Glycine composed anode-electrolyte interphase induced Zn(002) deposition for highly reversible zinc anode. *Chem. Eng. J.* **2024**, *480*, 148040. [CrossRef]
20. Yang, B.; Liu, B.; Luo, Z.; Yu, H.; Yin, F. Design and control the selection of martensitic variant to simultaneously improve strength and toughness of low-carbon martensitic steel. *Mater. Charact.* **2024**, *218*, 114540. [CrossRef]
21. Feitosa, A.; Ribamar, G.; Escobar, J.; Sonkusare, R.; Boll, T.; Coury, F.; Ávila, J.; Oliveira, J.; Padilha, A. Precipitation and reverted austenite formation in maraging 350 steel: Competition or cooperation? *Acta Mater.* **2024**, *270*, 119865. [CrossRef]
22. Ahluwalia, R.; Mikula, J.; Zeng, Y.; Laskowski, R.; Bai, K.; Vastola, G.; Zhang, Y.-W. Phase field study of heat treatment and strengthening in maraging steels. *Comput. Mater. Sci.* **2024**, *231*, 112590. [CrossRef]
23. Galindo-Nava, E.; Rainforth, W.; Rivera-Díaz-del-Castillo, P.E.J. Predicting microstructure and strength of maraging steels: Elemental optimisation. *Acta Mater.* **2016**, *117*, 270–285. [CrossRef]
24. Kapoor, M.; Isheim, D.; Ghosh, G.; Vaynman, S.; Fine, M.E.; Chung, Y.-W. Aging characteristics and mechanical properties of 1600MPa body-centered cubic Cu and B2-NiAl precipitation-strengthened ferritic steel. *Acta Mater.* **2014**, *73*, 56–74. [CrossRef]
25. Niu, M.; Zhou, G.; Wang, W.; Shahzad, M.B.; Shan, Y.; Yang, K. Precipitate evolution and strengthening behavior during aging process in a 2.5 GPa grade maraging steel. *Acta Mater.* **2019**, *179*, 296–307. [CrossRef]
26. Mei, X.; Yan, Y.; Fu, H.; Gao, X.; Huang, S.; Qiao, L. Effect of aging temperature on microstructure evolution and strengthening behavior of L-PBF 18Ni(300) maraging steel. *Addit. Manuf.* **2022**, *58*, 103071. [CrossRef]
27. Zhou, X.; Jia, C.; Mi, P.; Zhang, H.; Yan, W.; Wang, W.; Sun, M.; van der Zwaag, S.; Rong, L. Cyclic quenching treatment doubles the Charpy V-notch impact energy of a 2.3 GPa maraging steel. *J. Mater. Sci. Technol.* **2025**, *209*, 311–328. [CrossRef]
28. Vasilyev, D. Structural, elastic and thermal properties of Laves phase Fe2Mo from first principles. *J. Phys. Conf. Ser.* **2021**, *1942*, 012017. [CrossRef]
29. Yu, W.X.; Liu, B.X.; He, J.N.; Chen, C.X.; Fang, W.; Yin, F.X. Microstructure characteristics, strengthening and toughening mechanism of rolled and aged multilayer TWIP/maraging steels. *Mater. Sci. Eng. A* **2019**, *767*, 138426. [CrossRef]

30. Zhang, H.; Ji, X.; Ma, D.; Tong, M.; Wang, T.; Xu, B.; Sun, M.; Li, D. Effect of aging temperature on the austenite reversion and mechanical properties of a Fe–10Cr–10Ni cryogenic maraging steel. *J. Mater. Res. Technol.* **2021**, *11*, 98–111. [CrossRef]
31. Shamsujjoha, M. Evolution of microstructures, dislocation density and arrangement during deformation of low carbon lath martensitic steels. *Mater. Sci. Eng. A* **2020**, *776*, 139039. [CrossRef]
32. Rack, H.J. Age hardening-grain size relationships in 18Ni maraging steels. *Mater. Sci. Eng.* **1978**, *34*, 263–270. [CrossRef]

*materials*

MDPI

*Article*

# Machine Learning-Based Multi-Objective Composition Optimization of High-Nitrogen Austenitic Stainless Steels

Yinghu Wang [1,2], Long Chen [3], Limei Cheng [2], Enuo Wang [2], Zhendong Sheng [2] and Ligang Zhang [3,*]

1 National Center for Materials Service Safety, University of Science and Technology Beijing, Beijing 100083, China; hihihowareyou@163.com
2 Chengdu Institute of Advanced Metallic Material Technology and Industry Co., Ltd., Chengdu 610300, China; chenglimei_ustb@163.com (L.C.); panpan_wen@163.com (E.W.); shengzhendong@gmail.com (Z.S.)
3 School of Materials Science and Engineering, Central South University, Changsha 410083, China; cllong@csu.edu.cn
* Correspondence: ligangzhang@csu.edu.cn

**Abstract**

High-nitrogen austenitic stainless steels (HNASS) require compositional strategies that simultaneously maximize corrosion resistance and microstructural stability while suppressing delta (δ) ferrite and deleterious precipitates. Here, an explainable multi-objective design workflow is developed that couples thermodynamic descriptors from the Calculation of Phase Diagrams (CALPHAD) approach—using both equilibrium and Scheil solidification calculations—with machine learning surrogate models, random forest (RF) and Extreme Gradient Boosting (XGBoost), trained on 60,480 compositions in the Fe–C–N–Cr–Mn–Mo–Ni–Si space. The physics-informed feature set comprises phase fractions; transformation and precipitation temperatures for δ-ferrite, chromium nitride ($Cr_2N$), sigma (σ) phase and $M_{23}C_6$ carbides; liquidus and solidus temperatures; and the pitting-resistance equivalent number (PREN). The RF model achieves consistently low prediction errors, with a PREN root-mean-square error (RMSE) of ≈0.004, and exhibits strong generalization. Shapley additive explanations (SHAP) reveal metallurgically consistent trends: increasing nitrogen (N) suppresses δ-ferrite and promotes $Cr_2N$; carbon (C) promotes $M_{23}C_6$; molybdenum (Mo) promotes the σ-phase; and C and silicon (Si) widen the freezing range. Using the trained surrogate as the objective evaluator, the non-dominated sorting genetic algorithm III (NSGA-III) builds Pareto fronts that minimize the δ-ferrite range, $Cr_2N$, σ-phase, $M_{23}C_6$ and the freezing range (ΔT) while maximizing PREN. The Technique for Order Preference by Similarity to Ideal Solution (TOPSIS) is then applied to rank the Pareto-optimal candidates and to select compositions that combine elevated PREN with controlled precipitation windows. This workflow is efficient, reproducible and interpretable and provides actionable composition candidates together with a transferable methodology for data-driven stainless steel design.

**Keywords:** machine learning; high-nitrogen austenitic stainless steel; multi-objective genetic optimization; thermodynamic calculation of phase diagrams

## 1. Introduction

High-nitrogen austenitic stainless steels (HNASS) have emerged as a new generation of high-performance structural materials owing to their outstanding combination of corrosion resistance and mechanical properties in demanding service environments [1–4]. With a stable and homogeneous austenitic microstructure, HNASS are widely used in protective armor, shipbuilding, marine engineering and the chemical industry [5–10]. The solid-solution strengthening provided by nitrogen significantly increases the yield strength and hardness of HNASS. Moreover, nitrogen stabilizes the austenitic phase and enhances resistance to pitting and crevice corrosion [11–13], making these alloys particularly attractive for structural components that must withstand both high stresses and aggressive media. Benefiting from this property portfolio, HNASS are promising alternatives to conventional austenitic stainless steels, especially for applications under extreme conditions such as high pressure, elevated temperature and highly corrosive solutions. Therefore, systematic investigation of compositional design, microstructural control and performance evolution in HNASS is of considerable scientific and engineering significance. However, the limited solubility of nitrogen in molten steel, the propensity for nitride formation during solidification and the difficulty of precisely controlling nitrogen content during melting or additive manufacturing remain key challenges that hinder large-scale industrial deployment [14].

Although nitrogen is a minor alloying element, its influence on the tensile behavior [15–17] and fatigue performance [18–20] of steels has been extensively investigated. Nitrogen markedly increases fatigue life by promoting planar slip and inducing cyclic softening in the early stages of damage, thereby enhancing fatigue strength and reducing plastic strain accumulation; together, these effects delay crack initiation and retard subsequent crack growth [21–23]. Nitrogen additions also stabilize austenite and broaden the composition and temperature ranges of the austenitic phase, thereby suppressing ferrite formation [24–26]. In addition, nitrogen refines the austenitic grain structure [27–29] and improves the mechanical properties of austenitic stainless steels. From a corrosion perspective, nitrogen enhances localized corrosion resistance—particularly resistance to pitting and intergranular attack. For example, Gao et al. [30] examined the effect of nitrogen content on the composition, structure and protective performance of passive films in HNASS exposed to 0.5 mol $L^{-1}$ NaCl. Alloys with higher nitrogen contents exhibited a greater fraction of low-angle grain boundaries, lower corrosion current density and a reduced corrosion rate. The formation of stable oxides such as $Cr_2O_3$ in the passive film facilitates passivation and repassivation, underpinning the superior corrosion resistance of HNASS. Despite these advantages over conventional austenitic grades, the rate-dependent mechanical response of HNASS and the associated microstructural evolution remain less explored. Zhang et al. [31] systematically investigated the dynamic mechanical properties and microstructural evolution of HNASS over a wide strain-rate range, providing important insights but also highlighting the need for further mechanistic studies.

Despite these favorable properties, the compositional design of HNASS still faces significant challenges, primarily due to the complex coupling among nitrogen solubility, phase stability and corrosion resistance. Traditional trial-and-error alloy development and single-objective optimization are inefficient and often fail to capture nonlinear interactions among multiple alloying elements. Such empirical design approaches rely solely on experimental data and tend to overlook the underlying physical and chemical mechanisms. Furthermore, repeated alloy synthesis and testing make the design process time-consuming and costly.

To address these challenges, researchers have increasingly combined thermodynamic calculations, alloy design and data-driven optimization to tailor the composition and processing parameters of HNASS. Thermodynamic modeling based on the CALPHAD approach enables accurate prediction of phase stability, solubility limits and precipitation behavior in Fe–Cr–Mn–N–C–Si systems [32–35], and these predictions have been corroborated by experimental validations of phase equilibria and nitrogen solubility [36–38]. In parallel, machine learning (ML) has emerged as a powerful tool for materials optimization, efficiently exploring high-dimensional design spaces [39–42]. ML-assisted CALPHAD frameworks have been successfully applied to optimize microstructures and mechanical properties in steels, superalloys and titanium alloys, markedly improving prediction accuracy and computational efficiency [43–45]. In the context of high-nitrogen austenitic systems, Khan et al. [46] employed an extremely randomized trees (ETR) model to predict the yield strength and ultimate tensile strength of Cr–Mn–N alloys. Ahmadi et al. [47] developed a back-propagation artificial neural network (BP-ANN) to predict the high-temperature flow stress of API 5CT-L80 steel, achieving a high correlation coefficient and demonstrating the strong predictive capability of the BP-ANN model. Similarly, Wang et al. [48] constructed processing maps for P550 steel using a data-driven approach, in which flow-stress data generated by an ANN were used to build the maps and were subsequently validated experimentally, confirming the reliability of ANN-based process design.

Nevertheless, several critical scientific issues remain unresolved: (i) discrepancies between thermodynamic predictions and experimental observations under non-equilibrium solidification and rapid cooling; (ii) largely empirical selection of features when constructing composition–property relationships; (iii) the predominance of single-objective machine learning optimization that overlooks trade-offs between microstructural stability and corrosion resistance; and (iv) limited transferability and generalization of current models across alloy systems and processing routes. These limitations hinder quantitative and accurate compositional design of HNASS for engineering applications.

In view of the above, this study proposes an explainable, multi-objective alloy-design framework that integrates CALPHAD-derived thermodynamic data with machine learning algorithms to achieve composition–property optimization for HNASS. Thermodynamic descriptors—including liquidus and solidus temperatures, phase-transformation ranges and Gibbs energies—are extracted from the Fe–Cr–Mn–Ni–N–C–Si system via CALPHAD simulations. Feature-selection methods are then employed to identify the key factors governing microstructural stability and corrosion resistance. Subsequently, the trained surrogate is coupled with a multi-objective genetic algorithm, specifically the non-dominated sorting genetic algorithm III (NSGA-III) [49], to generate Pareto-optimal compositions that balance phase-stability descriptors and corrosion resistance.

This framework establishes a hybrid, theory- and data-driven pathway for intelligent compositional design of HNASS and provides a transferable methodology for developing next-generation stainless steels with superior performance and processability.

## 2. Materials and Methods

### 2.1. HTC Calculation

Thermodynamic calculations were performed within the CALPHAD framework using Thermo-Calc® 2021b with the TCFE11 database, which covers the relevant binary and multicomponent systems over the composition and temperature ranges pertinent to the target HNASS. The alloy design sought to balance mechanical stability and corrosion resistance while preserving austenite stability. Manganese (Mn) and nitrogen (N) were introduced to partially replace nickel (Ni), thereby stabilizing the austenitic phase and enhancing mechanical performance. Candidate compositions were selected with reference to the literature and commercial alloy specifications. Figure 1 summarizes the overall alloy composition design strategy and the associated thermodynamic–data-driven workflow.

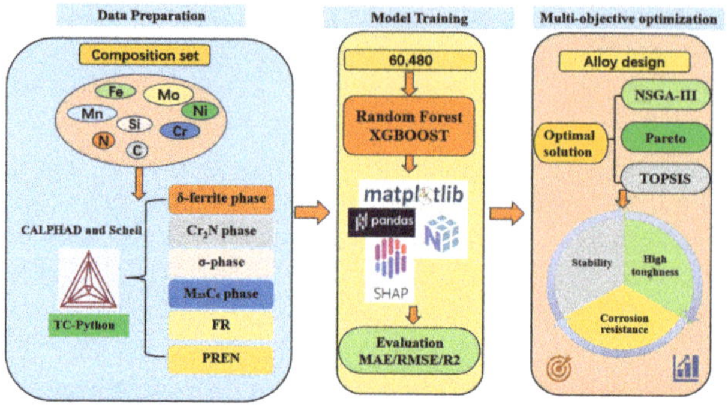


**Figure 1.** Alloy composition design strategy.

First, the compositional domain was defined from commercially available HNASS summarized in Ref. [50]. Around these baseline alloys, the ranges of C, N, Cr, Mn, Mo, Ni and Si were extrapolated to construct an extended design space (Table 1). Each element was discretized into a finite number of levels (e.g., six levels for N, five for Mo, six for Cr and Ni, seven for Mn, two for Si and four for C), and all combinations within the specified ranges were enumerated, yielding 60,480 candidate compositions. Subsequently, thermodynamic equilibrium calculations and Scheil solidification simulations were performed to obtain, for each composition, phase fractions, transformation/precipitation temperatures and solid–liquid boundaries (liquidus and solidus).

**Table 1.** Parameters used for the development of the compositional space in this study.

| Elements (wt.%) | Elemental Content | Step Length |
|---|---|---|
| **N (0.55–0.80)** | 0.55/0.60/0.65/0.70/0.75/0.80 | 0.05 |
| **Mo (1.8–2.6)** | 1.8/2.0/2.2/2.4/2.6 | 0.2 |
| **Cr (18–19)** | 18.0/18.2/18.4/18.6/18.8/19.0 | 0.2 |
| **Ni (3.5–4.5)** | 3.5/3.7/3.9/4.1/4.3/4.5 | 0.2 |
| **Mn (19.5–22.5)** | 19.5/20.0/20.5/21.0/21.5/22.0/22.5 | 0.5 |
| **Si (0.2–0.4)** | 0.2/0.4 | 0.2 |
| **C (0.01–0.04)** | 0.01/0.02/0.03/0.04 | 0.1 |
| **Total of Alloys' Candidates** | the total number of combinations: $6 \times 5 \times 6 \times 6 \times 7 \times 2 \times 4 = 60{,}480$ | **60,480** |

The precipitation of sigma ($\sigma$) phase—a Cr- and Mo-rich intermetallic compound—significantly deteriorates the toughness of high-nitrogen stainless steels, while excessive carbide formation and the presence of delta ($\delta$) ferrite also impair corrosion resistance and mechanical properties. Therefore, the following thermodynamic criteria were imposed in the present alloy design: (i) a wide single-phase austenite field must be present; (ii) the stability windows of $\sigma$ and carbide phases should be restricted—no $\sigma$ at 750 °C and no carbides above 960 °C; and (iii) the $\delta$-ferrite temperature range above 900 °C should be minimized. These thresholds were defined with reference to the equilibrium phase diagram of HNASS. This grid-based construction ensures that the initial sampling is uniform across the prescribed composition ranges, and any subsequent biases arise only from the thermodynamic and PREN-based filters rather than from the sampling procedure itself.

To further enhance the pitting resistance of high-nitrogen austenitic stainless steels and refine the candidate set, the pitting resistance equivalent number (PREN) was used as an additional screening descriptor. PREN accounts for the beneficial contributions of Cr and Mo, and its modified form $PREN_{\text{mod}}$ = Cr + 3.3Mo + 16N [51] explicitly incorporates the strong effect of nitrogen. Compositions with higher PREN values were preferentially retained for subsequent modeling and optimization.

Figure 2 summarizes the distributions of six descriptors. The δ-ferrite-related temperatures are strongly right-skewed, with a primary mode at 50–120 °C. The freezing range is unimodal, peaking at ≈180–220 °C with slight right skew. $Cr_2N$ precipitation temperatures cluster narrowly between 1000 and 1035 °C. The σ-phase sensitivity window centers at 960–970 °C and is approximately normal. $M_{23}C_6$ exhibits a clearly multimodal pattern with secondary modes near 780 °C and 840 °C. PREN is near-normally distributed and concentrated at 36–38 (dimensionless). Overall, the set displays both skewness and multimodality.

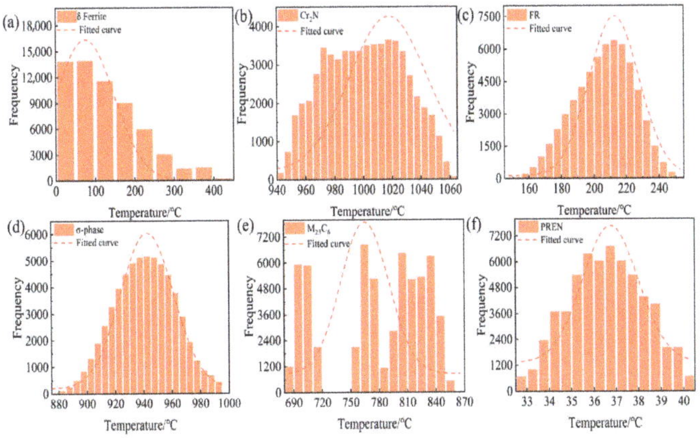


**Figure 2.** Histograms of the six target descriptors for the 60,480 candidate high-nitrogen austenitic stainless steel compositions: (**a**) δ-ferrite temperature range, (**b**) $Cr_2N$ precipitation/dissolution temperature, (**c**) freezing range ΔT, (**d**) σ-phase precipitation temperature, (**e**) $M_{23}C_6$ precipitation temperature and (**f**) PREN. Bars represent the frequency of CALPHAD-predicted values, and the dashed curves indicate the corresponding fitted probability density distributions.

These characteristics motivate the use of distribution-robust, tree-based models (e.g., random forest (RF) and XGBoost version 2.1.4), together with stratified sampling based on descriptor bins and, where appropriate, monotonic transformations (e.g., the Yeo–Johnson transformation) for the δ-ferrite and $M_{23}C_6$ targets. For downstream optimization, the objectives are defined to minimize the freezing range ΔT(FR) and the phase-formation temperatures of δ-ferrite, $Cr_2N$, σ-phase and $M_{23}C_6$ while maintaining a high PREN and avoiding the sensitivity windows associated with σ, $Cr_2N$ and $M_{23}C_6$.

## 2.2. Machine Learning Model

To accurately characterize the complex, nonlinear relationships among chemical composition, thermodynamic descriptors and performance metrics (e.g., corrosion resistance and phase stability), we employ two supervised learning algorithms—RF and Extreme Gradient Boosting (XGBoost)—to model data generated by thermodynamic simulations. RF is an ensemble method that constructs multiple decorrelated decision trees and averages their predictions, thereby improving generalization and mitigating overfitting, particularly in the presence of strong nonlinearities and high-dimensional feature interactions. XGBoost, a gradient-boosted decision tree algorithm, builds trees sequentially to minimize a regularized objective; its shrinkage, subsampling, and L1/L2 penalties provide effective bias–variance control and state-of-the-art predictive accuracy on tabular, heterogeneous data.

XGBoost is an efficient implementation of the gradient boosting decision tree (GBDT) framework. It trains an ensemble of weak learners sequentially by following the (approximate) negative gradient of a specified loss function, with each tree fitting the residuals left by the current ensemble. XGBoost delivers high predictive accuracy and training efficiency on large-scale datasets with sparse and nonlinearly coupled features. Its regularization (L1/L2 penalties), shrinkage (learning rate) and row/column subsampling—together with robust handling of missing values—mitigate overfitting and enhance generalization.

RF and Extreme Gradient Boosting (XGBoost) are complementary tree-based ensemble learners: RF implements bootstrap aggregation (bagging) of decorrelated decision trees, whereas XGBoost implements gradient-boosted decision trees (GBDT) [52–54]. Accordingly, we employ RF and XGBoost to train predictive models on the thermodynamics-derived dataset and to evaluate their accuracy and robustness.

### 2.3. Model Performance Metrics

To evaluate model performance, we used the root-mean-square error (*RMSE*), mean absolute error (*MAE*), coefficient of determination ($R^2$) and Pearson correlation coefficient ($r$) as performance metrics, computed as Equations (1)–(4) [55,56].

$$RMSE = \sqrt{\frac{1}{n}\sum_{i=1}^{n}(y_i - x_i)^2} \tag{1}$$

$$MAE = \frac{1}{n}\sum_{i=1}^{n}|y_i - x_i| \tag{2}$$

$$R^2 = 1 - \frac{\sum_{i=1}^{n}(y_i - x_i)^2}{\sum_{i=1}^{n}(\bar{y} - x_i)^2} \tag{3}$$

$$r = \frac{\sum_{i=1}^{n}(x_i - \bar{x})(y_i - \bar{y})}{\sqrt{\sum_{i=1}^{n}(x_i - \bar{x})^2}\sqrt{\sum_{i=1}^{n}(y_i - \bar{y})^2}} \tag{4}$$

where $x_i$ denotes the CALPHAD-derived target value (e.g., transformation temperature or PREN) for the $i$-th composition, $\bar{x}$ is the arithmetic mean of all CALPHAD target values in the evaluation set, $y_i$ denotes the corresponding value predicted by the surrogate model, $\bar{y}$ is the mean of all predicted values, and $n$ is the total number of samples in the evaluation set.

## 3. Results and Discussion

### 3.1. Correlation Analysis

In this study, six scalar objectives—δ-ferrite temperature range, $Cr_2N$ precipitation/dissolution temperature, σ-phase precipitation temperature, $M_{23}C_6$ precipitation temperature, PREN and the freezing range ΔT—are considered as functions of a common set of compositional features (C, N, Cr, Mn, Mo, Ni, Si). Pearson correlation analysis reveals several strong linear relationships (Figure 3). Carbon exhibits a very strong positive correlation with the $M_{23}C_6$ precipitation temperature ($r \approx 0.97$) and a moderate positive correlation with the freezing range ($r \approx 0.52$), indicating that the carbide-forming tendency is aligned with both precipitation temperature and the liquidus–solidus span. Molybdenum correlates strongly with the σ-phase precipitation temperature ($r \approx 0.79$) and moderately with PREN ($r \approx 0.55$), whereas chromium shows only weak-to-moderate correlations ($r \approx 0.28$–$0.29$) with selected transformation temperatures, including the σ-phase precipitation temperature and the δ-ferrite range. Nitrogen displays the opposite trend: it is strongly positively correlated with the $Cr_2N$ precipitation/dissolution temperature ($r \approx 0.90$) and strongly negatively correlated with the δ-ferrite range ($r \approx -0.85$); these two targets are themselves strongly negatively correlated ($r \approx -0.86$), reflecting the antagonism between nitride formation and ferrite stability. PREN correlates strongly with N ($r \approx 0.81$), moderately negatively with the δ-ferrite range ($r \approx -0.51$), and moderately positively with

Mo (r ≈ 0.55). Nickel shows near-zero first-order (linear) correlations with most targets, implying limited marginal effects within the studied domain. Overall, the correlation matrix exhibits a modular structure: (i) a positively correlated C–$M_{23}C_6$–Si–ΔT cluster; (ii) an N–$Cr_2N$ cluster that is negatively associated with the δ-ferrite range; and (iii) a Mo-centered cluster dominated by the σ-phase precipitation temperature.

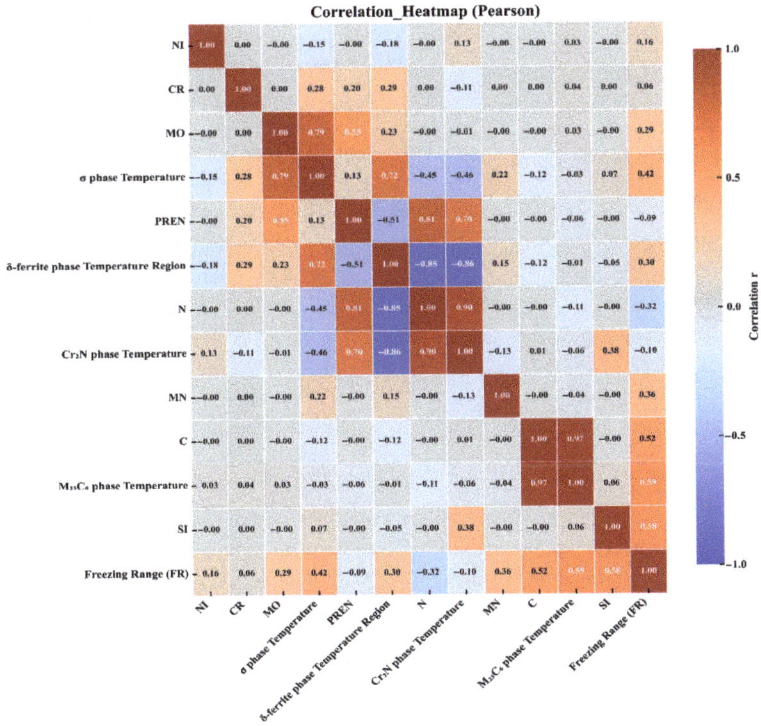


**Figure 3.** Correlation plot between molecular descriptors.

It should be emphasized that this figure is based on Pearson correlation (Figure 3), which measures first-order linear relationships and cannot be used for causal inference; extreme values and nonlinear relationships may weaken or obscure correlation strength. Since clustering used |r| as the similarity measure, variables grouped together represent "high correlation" rather than "same direction," and interpretation should also refer to the color (sign); Feature Description (Table 2). In this work, the input feature vector is restricted to the seven elemental mass fractions (C, N, Cr, Mn, Mo, Ni and Si), while the "physics-informed" aspect resides in the manual selection of thermodynamic target descriptors—δ-ferrite temperature range, $Cr_2N$, σ-phase and $M_{23}C_6$ precipitation temperatures, the freezing range and PREN—based on established metallurgical knowledge of phase stability and corrosion resistance in HNASS. This choice was subsequently validated by the Pearson correlation and SHAP analyses in Sections 3.1 and 3.3. As a sensitivity check, surrogate models retrained on reduced feature sets that excluded consistently low-importance elements exhibited only minor changes in $R^2$ and RMSE, indicating that the predictions are not unduly sensitive to these weak contributors.

**Table 2.** Feature description.

| Name | Description |
|---|---|
| δ-ferrite phase temperature | Ferritic phase that controls the solidification mode and strongly affects toughness and hot-cracking susceptibility in high-nitrogen austenitic stainless steels. |
| $Cr_2N$ phase temperature | Chromium nitride precipitate that consumes Cr and N from austenite, thereby degrading pitting resistance and lowering toughness, especially along grain boundaries. |
| σ-phase temperature | Cr- and Mo-rich intermetallic phase that causes severe embrittlement and depletes Cr and Mo from the matrix, reducing both toughness and localized corrosion resistance. |
| $M_{23}C_6$ phase temperature | Grain-boundary carbide that induces sensitization through local Cr depletion and simultaneously alters grain-boundary strength and creep resistance. |
| Freezing range | Temperature interval between the liquidus and the solidus; a larger freezing range promotes microsegregation, hinders feeding and increases the risk of solidification cracking. |
| PREN | Empirical pitting resistance equivalent number that quantifies resistance to localized corrosion in chloride-containing environments, with higher values indicating better performance. |

### *3.2. Evaluation of Different Models*

This study adopts a multi-target regression strategy in which six scalar objectives—δ-ferrite temperature range, $Cr_2N$ precipitation temperature, σ-phase precipitation temperature, $M_{23}C_6$ precipitation temperature, PREN and the freezing range $\Delta T$—are each modeled as functions of a common set of compositional features (C, N, Cr, Mn, Mo, Ni, Si). Tree-based ensemble learners, namely random forest (RF) and Extreme Gradient Boosting (XGBoost), are used as the core predictors. At the data-processing stage, any missing or non-finite feature values were imputed with the corresponding feature-wise mean. At the modeling stage, random seeds were fixed, the data were split into training and testing sets with an 80–20 ratio, and a one-model-per-target strategy was adopted to fit each objective separately. To obtain robust hyperparameters, a genetic algorithm was employed to optimize the XGBoost model by minimizing the three-fold cross-validation mean squared error; L1 and L2 regularizations (reg_alpha, reg_lambda), together with learning-rate shrinkage and subsampling, were used to mitigate overfitting. RF served as a parallel baseline and cross-validated comparator to assess the bias–variance trade-off. After training, RMSE and $R^2$ were computed on both the training and testing sets to quantify goodness of fit and generalization. The trained models and their evaluation outputs were stored for reproducibility and for subsequent interpretation and optimization. Overall, the pipeline combines cross-validation-guided evolutionary hyperparameter tuning with ensemble learning under reproducible settings, yielding accurate and stable predictors for all six targets.

Under the same data-cleaning and train–test splitting protocol, the random forest (RF) model consistently outperforms XGBoost across all six regression targets. Tables 3 and 4 summarize the performance metrics of the two machine learning models. For both the temperature-related objectives and the alloy corrosion-resistance indicator PREN, RF achieves lower RMSE and higher $R^2$, with very similar training and testing scores, indicating good generalization. For example, the test $R^2$ for the δ-ferrite precipitation temperature increases from 0.9113 with XGBoost to 0.9804 with RF, and the freezing range, $Cr_2N$ precipitation temperature and σ-phase precipitation temperature also show systematic improvements. For PREN, the RF model is nearly unbiased (RMSE ≈ 0.004, $R^2$ ≈ 0.999), whereas XGBoost still exhibits noticeable errors (RMSE ≈ 0.3086, $R^2$ ≈ 0.9659). These results suggest that, for the present feature–response relationships and hyperparameter choices, the bagging-based tree ensemble captures nonlinearities and interactions more effectively, while the gradient-boosting model would require further tuning (e.g., more weak learners, relaxed regularization and a broader hyperparameter search) to reach comparable performance.

**Table 3.** XGBoost model prediction metrics for each feature.

| Type | Dataset Type | $R^2$ | MAE | RMSE |
|---|---|---|---|---|
| δ-ferrite phase temperature Region | Training set<br>Test set | 0.922<br>0.911 | 18.387<br>18.701 | 26.058<br>27.819 |
| Freezing range | Training set<br>Test set | 0.917<br>0.916 | 4.309<br>4.311 | 5.231<br>5.249 |
| $Cr_2N$ precipitation temperature | Training set<br>Test set | 0.953<br>0.929 | 4.904<br>4.931 | 5.972<br>6.024 |
| σ-phase temperature | Training set<br>Test set | 0.930<br>0.929 | 4.446<br>4.508 | 5.659<br>5.732 |
| $M_{23}C_6$ phase temperature | Training set<br>Test set | 0.974<br>0.974 | 6.792<br>6.758 | 8.204<br>8.164 |
| PREN | Training set<br>Test set | 0.966<br>0.965 | 0.254<br>0.254 | 0.313<br>0.309 |

**Table 4.** Random forest model prediction metrics for each feature.

| Type | Dataset Type | $R^2$ | MAE | RMSE |
|---|---|---|---|---|
| δ-ferrite phase temperature Region | Training set<br>Test set | 0.984<br>0.980 | 10.44<br>10.31 | 13.91<br>12.95 |
| Freezing range | Training set<br>Test set | 0.986<br>0.98. | 1.37<br>1.27 | 1.76<br>1.65 |
| $Cr_2N$ precipitation temperature | Training set<br>Test set | 0.999<br>0.998 | 0.98<br>0.979 | 1.27<br>1.23 |
| σ-phase temperature | Training set<br>Test set | 0.981<br>0.978 | 2.11<br>2.11 | 2.62<br>2.65 |
| $M_{23}C_6$ phase temperature | Training set<br>Test set | 0.974<br>0.974 | 1.08<br>1.09 | 1.43<br>1.41 |
| PREN | Training set<br>Test set | 0.999<br>0.999 | 0.003<br>0.003 | 0.004<br>0.004 |

To intuitively evaluate the fitting and generalization performance of the two types of models on each target, this paper plots prediction–true value comparison charts for RF and XGBoost (Figure 4, Figure 5), respectively. Each subplot uses equal-axis coordinates to display the predicted and true values of the test/training samples, with y = x as the reference baseline; a consistent Nature Publishing Group (NPG) color scheme is used to distinguish data subsets (circles for the training set, stars for the test set), with outer edge strokes to enhance recognizability. To improve readability of the scatter plots, points with absolute residuals exceeding a threshold of $|y - \hat{y}| > 3$ (in the respective physical units) are visually truncated. This heuristic choice corresponds to removing less than X% of the samples and serves only to reduce clutter; it does not affect the fitted models or the numerical error metrics reported in Tables 3 and 4. The results show that for all targets, the scatter points of RF are more tightly clustered along y = x, with highly consistent training and test distributions;

XGBoost shows more obvious dispersion and deviation in indicators such as δ-ferrite phase range precipitation temperature and FR. Combined with the aforementioned RMSE/$R^2$ metrics (for example, PREN under RF is almost unbiased, $R^2 \approx 1.000$; whereas XGBoost still has visible errors), it can be considered that bagging-based ensembles are more robust to the current feature–response relationships and noise structure, exhibiting stronger generalization consistency. It should be emphasized that these performance metrics quantify how accurately the surrogates reproduce the CALPHAD-derived thermodynamic descriptors, rather than experimental measurements. The models are therefore intended as numerical surrogates for the underlying thermodynamic calculations.

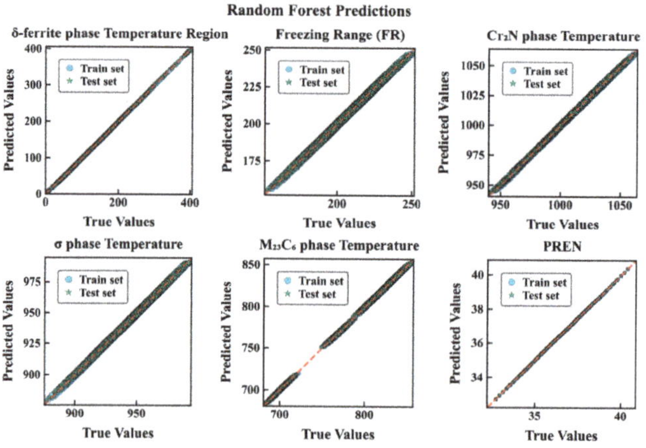


**Figure 4.** Predicted versus true values for six target properties using the RF regression model.

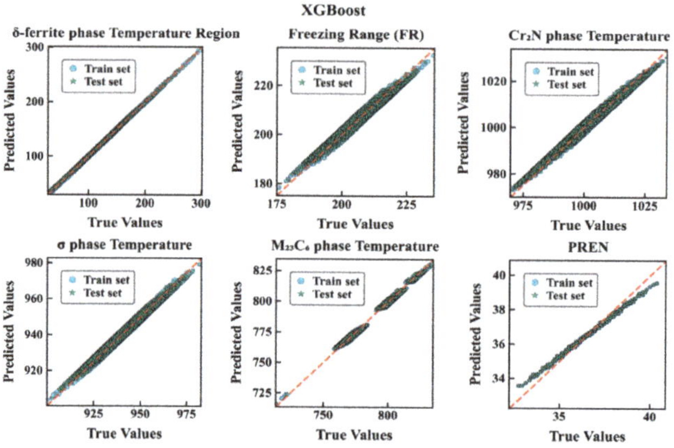


**Figure 5.** Predicted versus true values for six target properties using the XGBoost regression model.

### 3.3. SHAP Analysis

The SHAP (SHapley Additive exPlanations) value is a method for explaining model prediction results. It evaluates the importance of each feature by calculating the contribution of the feature to the model output. The essence of the SHAP value is a Shapley value, which originates from the concept of the Shapley value in game theory, used to measure each player's contribution in a game.

In machine learning, SHAP values are used to explain the importance contribution of the model's prediction results for each sample. Specifically, for each sample, the SHAP value tells us the degree to which each feature influences the model's prediction for that sample, thereby explaining the model's prediction results, helping to understand the model's decision-making process, and discovering hidden patterns and trends in the data.

SHAP values can be used to explain the prediction results of a RF model, and the summary plot is a commonly used visualization method. The summary plot can intuitively show the contribution of each feature to the model's prediction results and can also consider multiple features at the same time, helping us understand the model's decision-making process. SHAP values provide insights into how each feature affects prediction outcomes by capturing both positive and negative effects across different samples [57,58].

To gain a deeper understanding of the impact of different component elements on the prediction results of various performance indicators, this study calculated Shapley Additive Explanation values (SHAP values) based on the RF model and plotted SHAP summary plots for each target variable (Figure 6). The horizontal axis represents the SHAP values, reflecting the direction and magnitude of a feature's marginal contribution to the prediction output; the vertical axis represents the input features, and the color of the points indicates the feature value (blue for low values, red for high values). The greater the horizontal dispersion of the points, the stronger the influence of that feature on the model output.

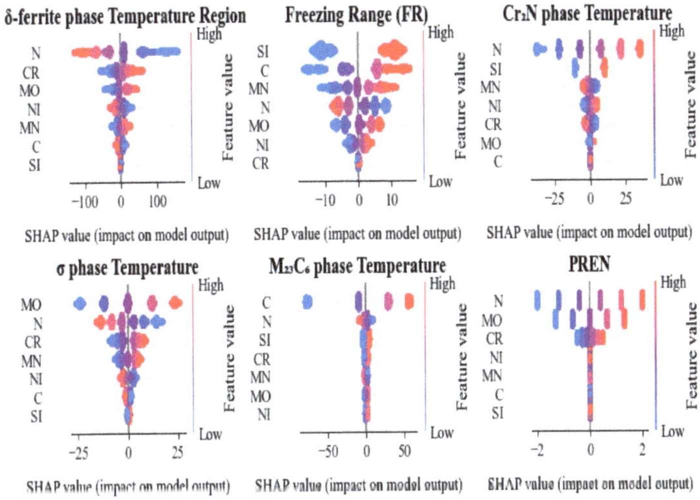


**Figure 6.** Model interpretation analyzed by SHAP.

First, in the prediction of the δ-ferrite phase temperature region, nitrogen (N) appears as the primary driving factor, with high N content showing a significant negative contribution to the predicted value, while chromium (Cr) and molybdenum (Mo) exhibit positive effects in certain ranges, indicating their important role in stabilizing phase transformation temperature.

As shown in Figure 7, RF feature importances identify physically consistent drivers for the six targets. The δ-ferrite temperature range is dominated by N, with Cr and Mo as secondary contributors and Ni and Mn minor; C and Si are negligible. The freezing range is governed mainly by Si and C, with smaller effects from Mn and Mo. The $Cr_2N$ precipitation temperature is driven primarily by N. The σ-phase temperature is governed by Mo. The $M_{23}C_6$ precipitation temperature is controlled by C. PREN depends strongly on N and Mo, with a weaker contribution from Cr. These rankings accord with established metallurgy—N stabilizes austenite and promotes nitrides, Mo promotes σ, C governs carbide formation, and C and Si widen the liquidus–solidus interval—they are consistent with our SHAP and permutation analyses.

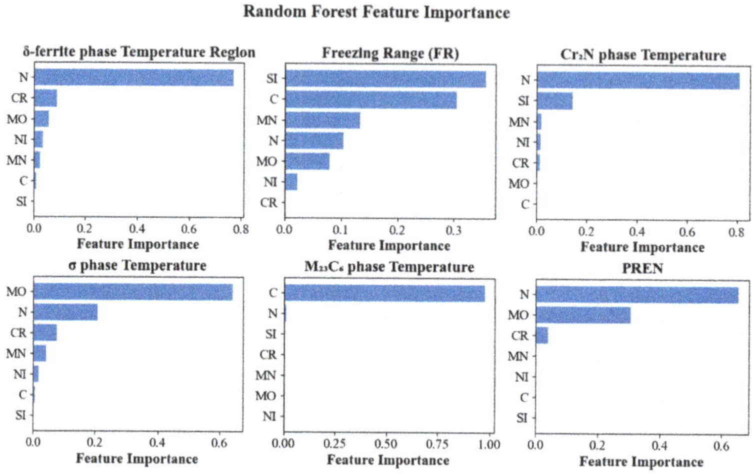


**Figure 7.** Random forest feature importance for six targets.

For the freezing range, silicon (Si) and carbon (C) emerge as the dominant predictors. High Si values are associated with negative SHAP contributions, indicating a narrowing of the freezing range, whereas high C values show positive SHAP contributions, indicating an expansion of the freezing range. Manganese (Mn) and nitrogen (N) provide moderate contributions. These results highlight the multi-element, synergistic regulation of solidification behavior and underscore the importance of considering coupled compositional effects when optimizing the freezing range.

For the $Cr_2N$ precipitation temperature, nitrogen (N) is the dominant predictor: higher N levels yield strongly positive SHAP values, indicating an increase in the predicted transformation temperature. In contrast, chromium (Cr) and molybdenum (Mo) exhibit negative contributions, lowering the predicted temperature within the examined composition range. This pattern suggests that elevated nitrogen activity stabilizes $Cr_2N$, whereas Cr and Mo tend to depress its transformation temperature under the present conditions.

For the prediction of the σ-phase precipitation temperature, molybdenum (Mo) is the most influential feature: higher Mo levels produce strongly positive effects, indicating an increase in the predicted σ-phase precipitation temperature. Chromium (Cr) and PREN-related descriptors also contribute positively, albeit to a lesser extent. These trends are consistent with the metallurgical understanding that Mo and Cr promote σ-phase formation within the examined composition range.

For $M_{23}C_6$ phase, carbon (C) is the most critical contributing factor, with high carbon levels strongly promoting the precipitation of $M_{23}C_6$-type carbides; meanwhile, high N tends to form nitrides, which suppress $M_{23}C_6$ precipitation, and other components have limited influence on $M_{23}C_6$. This conclusion aligns with the theory that carbide formation depends on the content of carbon and transition metal elements.

Finally, for PREN prediction, nitrogen (N), chromium (Cr) and molybdenum (Mo) are the dominant positive contributors, indicating that PREN is governed primarily by these corrosion-resistant elements. The remaining elements exert only minor effects, acting as weak modifiers over limited compositional ranges.

In summary, SHAP analyses across the six prediction tasks validate the reliability of the models and corroborate established alloy-design and phase-transformation theories, while revealing distinct, target-dependent elemental effects that differentiate the contributions of multiple elements to the predicted microstructural and performance metrics.

These SHAP patterns are consistent with the established metallurgical understanding of HNASS and allow each ML prediction to be decomposed into contributions from individual elements, thereby providing a physically interpretable link between composition, phase stability descriptors and corrosion resistance.

### 3.4. Decision Variables and Value Ranges

#### 3.4.1. Model Validation

To enable collaborative optimization of multiple performance indicators in alloy composition design, we formulate a six-objective optimization problem. The decision variables are the mass fractions of seven principal elements—C, N, Cr, Mn, Mo, Ni, and Si—within the prescribed composition bounds. The objective vector comprises the predicted values of six targets: the δ-ferrite temperature range, the freezing range, the $Cr_2N$ precipitation temperature, the σ-phase precipitation temperature, the $M_{23}C_6$ precipitation temperature and PREN. Consistent with design goals, the first five objectives are minimized and PREN is maximized. Each objective is evaluated by a surrogate model based on RF regression trained in the preceding section.

#### 3.4.2. Optimize Objective Function

The optimization objective consists of six performance indicators, denoted, respectively, as follows:

$$f(X) = (f1(X), f2(X), f3(X), f4(X), f5(X), f6(X)) \tag{5}$$

This multi-objective optimization design aims to simultaneously reduce the precipitation risk of harmful phases (σ-phase, $M_{23}C_6$, etc.) and significantly improve pitting resistance. The resulting alloy composition system corresponds to high-nitrogen, low-carbon super-austenitic/super-duplex stainless steel, capable of achieving excellent corrosion resistance and structural stability in high-chloride and highly corrosive environments, among others (Table 5).

**Table 5.** Performance metric range.

| Number | Name | Value |
|---|---|---|
| f1 | δ-ferrite phase temperature Region | Min |
| f2 | $Cr_2N$ phase temperature | Min |
| f3 | σ-phase temperature | Min |
| f4 | $M_{23}C_6$ phase temperature | Min |
| f5 | the freezing range | Min |
| f6 | PREN | Max |

The corresponding target orientation is set to [1,1,1,1,1,−1], where "1" indicates a minimization objective and "−1" indicates a maximization objective.

#### 3.4.3. Optimization Algorithms and Parameter Settings

In order to simultaneously optimize the six conflicting design objectives considered in this work—minimizing the δ-ferrite temperature range, the freezing range and the precipitation/dissolution temperatures of $Cr_2N$, σ-phase and $M_{23}C_6$, while maximizing PREN—the non-dominated sorting genetic algorithm III (NSGA-III) is employed to explore the composition space spanned by C, N, Cr, Mn, Mo, Ni and Si. NSGA-III is an improved version of the widely used NSGA-II algorithm and is specifically designed for many-objective optimization problems involving more than three objectives [59,60]. By combining classical non-dominated sorting with a set of predefined reference points in the objective space, NSGA-III preserves population diversity and yields a well-distributed approximation to the Pareto front even in six-dimensional objective space. These characteristics make NSGA-III particularly suitable for the present stainless steel design problem, where complex trade-offs between phase-stability descriptors and pitting-resistance must be captured quantitatively before applying the subsequent TOPSIS-based ranking.

As illustrated in Figure 8, the NSGA-III Pareto front is visualized using a parallel-coordinates plot in which all six objectives are scaled to [0,1] (lower values indicate better performance for the five temperature-related objectives, whereas higher values are preferred for PREN). Several consistent patterns can be observed. First, many trajectories combine a low δ-ferrite temperature range with a high PREN, indicating a favorable synergy between suppressing δ-ferrite and enhancing corrosion resistance. Second, high-PREN solutions frequently coincide with elevated $Cr_2N$ precipitation temperature and, to a lesser extent, higher σ-phase precipitation temperature, revealing an inherent trade-off between corrosion resistance and the tendency to form nitrides and intermetallics. Third, the $M_{23}C_6$ precipitation temperature can often be reduced concurrently with the σ-phase precipitation temperature, whereas the freezing range remains a bottleneck: solutions that strongly minimize the freezing range tend to exhibit only moderate PREN or an increase in at least one precipitation temperature. Overall, the Pareto front exhibits clear "knee" regions characterized by low δ-ferrite range, low-to-moderate σ and $M_{23}C_6$ precipitation temperatures, an acceptable freezing range and high PREN; these compromise solutions were subsequently prioritized in the TOPSIS-based ranking step.

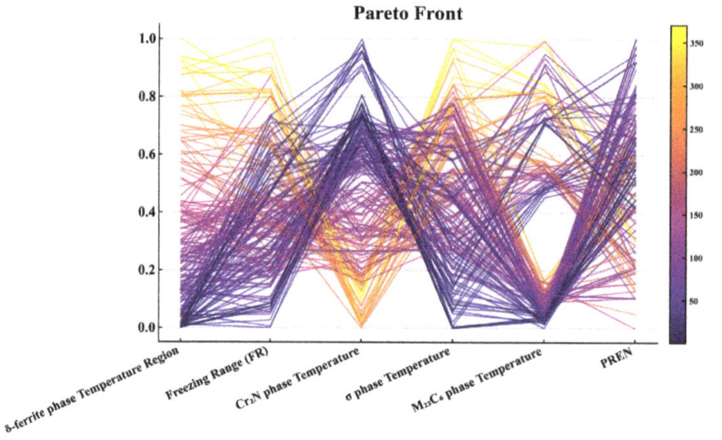


**Figure 8.** Parallel coordinate plot showing six optimized objectives from the MOEA, with red lines representing Pareto-optimal solutions.

### 3.5. Results Screening and Evaluation

After optimization, the non-dominated solutions form the Pareto front. To select representative candidates, we apply TOPSIS for comprehensive ranking. The six objective values for each solution are first normalized to a common scale and assigned equal weights of one. We then define an ideal best point with maximum PREN and minimum values for the five temperature-related objectives and an ideal worst point with the opposite attributes. For every solution, Euclidean distances to these two reference points are computed and converted into a relative closeness score, where a larger score indicates a better overall compromise. The highest-scoring solutions are retained for subsequent analysis and validation.

In this work, TOPSIS is applied to rank the Pareto-optimal solutions obtained from NSGA-III. First, all six objectives are normalized across the Pareto set to remove scale effects, with the five temperature-related descriptors treated as cost-type criteria (to be minimized) and PREN treated as a benefit-type criterion (to be maximized). The ideal best point in the normalized objective space is then constructed by taking the minimum values for all temperature-related objectives and the maximum value for PREN, whereas the ideal worst point is constructed by taking the maximum temperatures and the minimum PREN. For each candidate alloy, the Euclidean distances to the ideal best and ideal worst points are computed, and a relative closeness coefficient is obtained as the ratio between the distance to the worst point and the sum of both distances. A higher closeness coefficient therefore indicates a more favorable trade-off among the six objectives. In this first demonstration, equal weights are assigned to all objectives to avoid imposing subjective preferences; alternative weightings can be introduced in future work to emphasize specific design priorities (e.g., corrosion resistance versus phase-stability margins).

### 3.6. Optimization Results

The optimization results show that the TOPSIS relative closeness of the top ten candidates exceeds 0.82, indicating well-balanced performance across objectives. The top-ranked composition—C 0.01, N 0.80, Cr 18.0, Mn 19.5, Mo 1.80, Ni 3.90, Si 0.20 wt.%—achieves a closeness of 0.922 and yields the following predictions: δ-ferrite temperature range $\approx 0$ °C, freezing range $\approx 155.0$ °C, $Cr_2N$ precipitation temperature $\approx 1034.8$ °C, σ-phase precipitation temperature $\approx 884.3$ °C, $M_{23}C_6$ precipitation temperature $\approx 689.1$ °C and PREN $\approx 36.74$ (dimensionless). Other high-ranking solutions (ranks 2–5) present similarly credible trade-offs between PREN and the precipitation temperatures, confirming the effectiveness of NSGA-III combined with TOPSIS and the robustness of the surrogate-based optimization workflow.

In summary, multi-objective optimization followed by TOPSIS ranking produced a Pareto set within the prescribed bounds and identified composition candidates with superior trade-offs among objectives. These outcomes provide a quantitative basis for subsequent alloy design and performance tailoring.

Figure 9 presents the calculated equilibrium phase evolution for the optimized high-nitrogen austenitic stainless steel composition Fe–18.0Cr–19.5Mn–1.80Mo–3.90Ni–0.20Si–0.01C–0.80N (wt.%), obtained using the Calculation of Phase Diagrams (CALPHAD) approach with the TCFE13 database. The results show that austenite (γ) is thermodynamically stable over a broad interval of approximately 950–1400 °C, constituting nearly 100 wt.% of the solid prior to melting. The lack of an appreciable δ-ferrite field within this range indicates that the designed chemistry effectively stabilizes the austenitic matrix and suppresses δ-ferrite formation, which supports the targeted combination of high toughness and corrosion resistance.

At lower temperatures (<900 °C), small fractions of secondary phases—$M_2(C,N)$, $M_{23}C_6$ carbides and σ—are predicted. The $M_2(C,N)$ phase is stable over approximately 900–700 °C, consistent with the formation of Cr- and Mo-bearing carbo-nitrides that can contribute to precipitation strengthening. The σ-phase fraction remains limited (<5 wt.%), indicating a relatively low risk of brittle intermetallic embrittlement. A minor amount of Laves phase is also predicted below ~800 °C, which is typical for Mo-bearing austenitic steels. Crucially, all secondary phases dissolve completely above ~950 °C, enabling a homogeneous single-phase austenite during solution treatment and providing a favorable starting point for subsequent thermomechanical processing and mechanical performance.

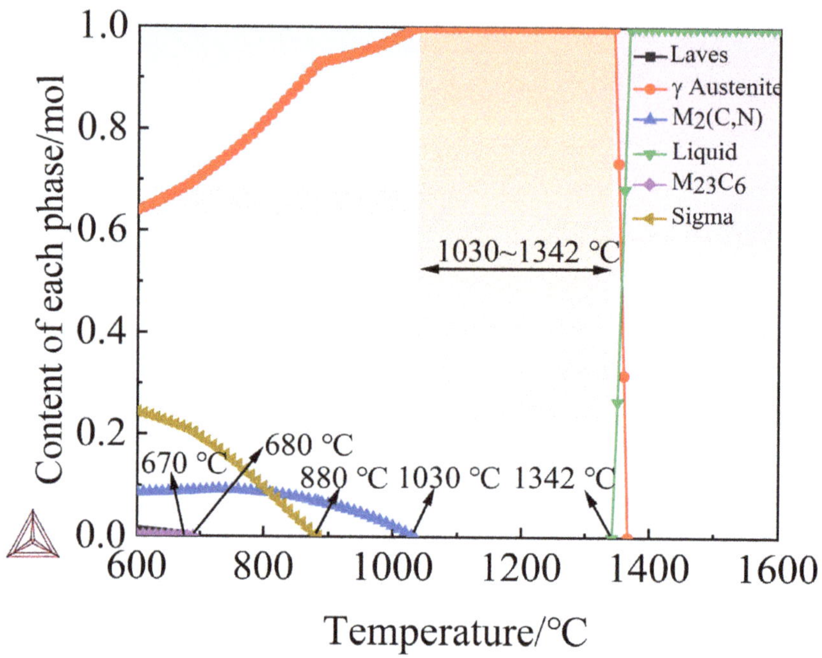


**Figure 9.** Equilibrium phase fractions of the designed high-nitrogen austenitic stainless steel (Fe–18.0Cr–19.5Mn–1.8Mo–3.9Ni–0.2Si–0.01C–0.8N wt.%) calculated using the CALPHAD method with the TCFE11 database.

Overall, thermodynamic simulations confirm that the machine learning-optimized composition achieves a balanced combination of austenite stability and controlled precipitation. This balance is essential for realizing the targeted synergy of high strength, ductility, and corrosion resistance in high-nitrogen austenitic stainless steels.

As shown in Figure 10, the Scheil–Gulliver non-equilibrium path of the high-nitrogen austenitic stainless steel exhibits primary austenite solidification (L $\rightarrow$ L + $\gamma$ $\rightarrow$ $\gamma$), which simplifies transformation sequences and promotes reproducible, fully austenitic microstructures. The well-defined liquidus and solidus provide a clear freezing range that can be directly used to set process windows (e.g., preheat level, linear energy, scan speed) and to calibrate thermal–metallurgical simulations. Austenite-first solidification also helps retain nitrogen in the matrix, supporting higher PREN and good toughness by avoiding sequestration of N into δ-ferrite. Moreover, the smooth, monotonic $f_s$–T evolution and distinct landmarks (nucleation onset, coherency/coalescence intervals) yield clean, ML-ready descriptors, facilitating data-driven multi-objective optimization that links composition, processing, and properties. Overall, the figure offers actionable guidance for process design while preserving the intrinsic advantages of HNASS.

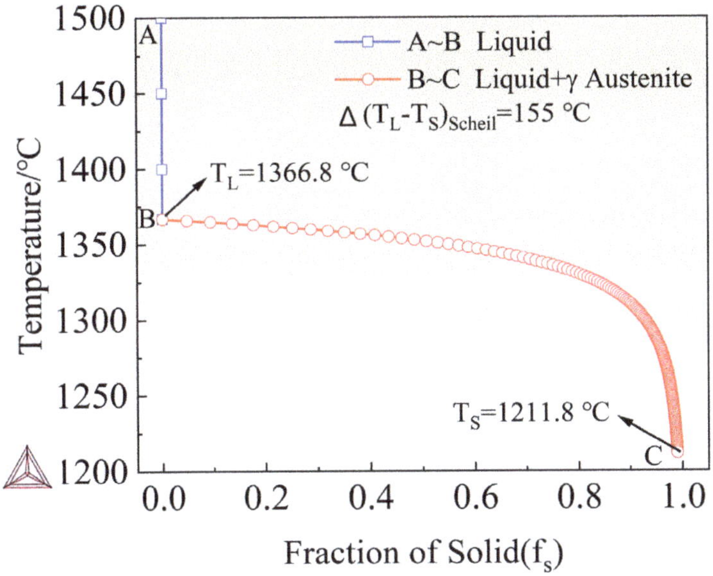


**Figure 10.** Scheil–Gulliver solidification path of HNASS. A–B: liquid; B–C: liquid + $\gamma T_L = 1366.8°C$, $T_S = 1211.8°C$; $\Delta T_{Scheil} = 155°C$.

## 4. Conclusions

We constructed a six-objective optimization framework in which the mass fractions of seven principal elements (C, N, Cr, Mn, Mo, Ni, Si) serve as decision variables. The objectives are to minimize the δ-ferrite temperature range, the freezing range, the $Cr_2N$ precipitation temperature, the σ-phase precipitation temperature and the $M_{23}C_6$ precipitation temperature, while maximizing PREN. Using a random forest surrogate trained on thermodynamics-derived data, the optimized composition reduces the propensity for harmful precipitates (notably σ and $M_{23}C_6$), narrows the freezing range and thereby mitigates segregation and hot-cracking susceptibility, improving microstructural stability during welding and casting. While maintaining mechanical properties, the optimized alloy exhibits a substantially increased PREN; the resulting pitting resistance approaches or exceeds that of super-austenitic and super-duplex stainless steels, indicating suitability for long-term service in chloride-rich and strongly corrosive environments such as marine engineering, seawater desalination and chemical processing. This multi-objective strategy provides a quantitative basis and a transferable methodology for the compositional design of high-performance, corrosion-resistant stainless steels. Although the RF model achieves very high $R^2$ values and low RMSE with respect to the CALPHAD data, this does not automatically guarantee agreement with experimental measurements. To mitigate overfitting at the surrogate level, we employed an 80–20 train–test split with fixed random seeds, three-fold cross-validation during hyperparameter optimization and controlled tree complexity. The close agreement between training and testing metrics (Tables 3 and 4) and the absence of systematic divergence in the parity plots (Figures 4 and 5) suggest that variance-driven overfitting is limited. Nevertheless, the overall predictive fidelity remains bound by the accuracy of the thermodynamic database, and experimental validation of the optimized compositions is required before industrial deployment. By construction, the six target descriptors in this work are outputs of equilibrium and Scheil CALPHAD calculations. The RF and XGBoost models are therefore trained to reproduce these thermodynamic quantities, and the high $R^2$ values together with the narrow residual distributions indicate that the surrogates closely match the CALPHAD predictions without obvious systematic deviations in specific temperature ranges. The discrepancy between the present predictions and actual experimental behavior is thus primarily governed by the fidelity of the underlying thermodynamic database. Future work will focus on benchmarking representative optimized compositions against experimental phase-boundary and precipitation data to quantify these deviations more precisely.

Practical manufacturability considerations are partially embedded in the present framework through the choice of composition ranges and thermodynamic filters: the ranges of C, N, Cr, Mn, Mo, Ni and Si are centered on commercial HNASS and reported alloys, and compositions with excessive δ-ferrite, σ-phase or high-temperature carbides are removed, which indirectly promotes weldability and microstructural robustness. However, explicit cost metrics and process-route-specific constraints (e.g., nitrogen solubility margins for pressurized melting or processability windows for additive manufacturing) are not included as separate objectives or constraints in the current NSGA-III formulation. Future extensions of the framework could incorporate such quantities directly, enabling truly manufacturing-aware multi-objective optimization.

In future work, first-principle calculations will be integrated with multi-source experimental datasets to further improve the predictive accuracy of the models. Building on this, systematic experimental validation and service-performance evaluation will be conducted, with emphasis on the synergistic effects of thermomechanical processing routes and heat-treatment schedules on the freezing range, microstructural stability, and corrosion behavior. In parallel, the multi-objective optimization framework will be extended to a broader set of stainless steel systems and other high-performance alloys, enabling coordinated improvements in mechanical properties, corrosion resistance, microstructural stability, and processability.

The present study is purely computational and does not yet include experimental validation of the optimized compositions. Planned future work will synthesize representative Pareto-optimal alloys, followed by microstructural and corrosion characterization, in order to benchmark both the CALPHAD predictions and the ML-based optimization against experimental behavior.

**Author Contributions:** Y.W.: Conceptualization, Methodology, High-Throughput Computation, Validation, Data Curation, Writing—Original Draft, Writing—Review and Editing. L.C. (Long Chen): Conceptualization, Methodology. L.C. (Limei Cheng): Investigation, Writing—Review and Editing. E.W.: Supervision, Project Administration. Z.S.: Methodology, Writing—Review and Editing. L.Z.: Methodology, Validation, Writing—Review and Editing. All authors have read and agreed to the published version of the manuscript.

**Funding:** This work was financially supported by the Joint Training Scholarship for PhD/Master's Students in Key Industrial Sectors (Grant No. 2025-RC04-00165-RC).

**Data Availability Statement:** The original contributions presented in this study are included in the article. Further inquiries can be directed to the corresponding author.

**Conflicts of Interest:** Authors Yinghu Wang, Limei Cheng, Enuo Wang, and Zhendong Sheng were employed by the company Chengdu Institute of Advanced Metallic Material Technology and Industry Co., Ltd. The remaining authors declare that the research was conducted in the absence of any commercial or financial relationships that could be construed as a potential conflict of interest.

## References


1. Wang, T.H.; Wang, J.; Bai, J.G.; Wang, S.J.; Chen, C.; Han, P.D. Effect of boron on dissolution and repairing behavior of passive film on S31254 super-austenitic stainless steel immersed in $H_2SO_4$ solution. *J. Iron Steel Res. Int.* **2022**, *29*, 1012–1025. [CrossRef]
2. Pujar, M.G.; Kamachi Mudali, U.; Singh, S.S. Electrochemical Noise Studies of the Effect of Nitrogen on Pitting Corrosion Resistance of High Nitrogen Austenitic Stainless Steels. *Corros. Sci.* **2011**, *53*, 4178–4186. [CrossRef]
3. Olsson, C.-O.A.; Landolt, D. Passive Films on Stainless Steels—Chemistry, Structure and Growth. *Electrochim. Acta* **2003**, *48*, 1093–1104. [CrossRef]
4. Wang, S.; Gao, Z.; Wu, G.; Mao, X. Titanium microalloying of steel: A review of its effects on processing, microstructure and mechanical properties. *J. Univ. Sci. Technol. Beijing* **2022**, *29*, 645–661. [CrossRef]
5. Qiao, Y.X.; Zheng, Z.B.; Yang, H.K.; Long, J.; Han, P.X. Recent progress in microstructural evolution, mechanical and corrosion properties of medium-Mn steel. *J. Iron Steel Res. Int.* **2023**, *30*, 1463–1476. [CrossRef]
6. Talha, M.; Behera, C.K.; Sinha, O.P. In-Vitro Long Term and Electrochemical Corrosion Resistance of Cold Deformed Nitrogen Containing Austenitic Stainless Steels in Simulated Body Fluid. *Mater. Sci. Eng. C* **2014**, *40*, 455–466. [CrossRef]
7. Poonguzhali, A.; Pujar, M.G.; Kamachi Mudali, U. Effect of Nitrogen and Sensitization on the Microstructure and Pitting Corrosion Behavior of AISI Type 316LN Stainless Steels. *J. Mater. Eng. Perform.* **2013**, *22*, 1170–1178. [CrossRef]
8. Li, H.; Jiang, Z.; Zhang, Z.; Cao, Y.; Yang, Y. Intergranular Corrosion Behavior of High Nitrogen Austenitic Stainless Steel. *Int. J. Miner. Metall. Mater.* **2009**, *16*, 654–660.
9. Erneman, J.; Schwind, M.; Liu, P.; Nilsson, J.-O.; Andrén, H.-O.; Ågren, J. Precipitation Reactions Caused by Nitrogen Uptake during Service at High Temperatures of a Niobium Stabilised Austenitic Stainless Steel. *Acta Mater.* **2004**, *52*, 4337–4350. [CrossRef]
10. Lang, Y.P.; Qu, H.P.; Chen, H.T.; Weng, Y.Q. Research progress and development tendency of nitrogen-alloyed austenitic stainless steels. *J. Iron Steel Res. Int.* **2015**, *22*, 91–98. [CrossRef]
11. Garcia-Cabezon, C.; Blanco, Y.; Rodriguez-Mendez, M.L.; Martin-Pedrosa, F. Characterization of porous nickel-free austenitic stainless steel prepared by mechanical alloying. *J. Alloys Compd.* **2017**, *716*, 46–55. [CrossRef]
12. Ningshen, S.; Mudali, U.K.; Mittal, V.K.; Khatak, H.S. Semiconducting and passive film properties of nitrogen-containing type 316LN stainless steels. *Corros. Sci.* **2007**, *49*, 481–496. [CrossRef]
13. Hu, L.; Ngai, T.; Peng, H.; Li, L.; Zhou, F.; Peng, Z. Microstructure and Properties of Porous High-N Ni-Free Austenitic Stainless Steel Fabricated by Powder Metallurgical Route. *Materials.* **2018**, *11*, 1058. [CrossRef] [PubMed]
14. Ha, H.Y.; Lee, T.H.; Oh, C.S.; Kim, S.J. Effects of combined addition of carbon and nitrogen on pitting corrosion behavior of Fe–18Cr–10Mn alloys. *Scr. Mater.* **2009**, *61*, 121–124. [CrossRef]
15. Moon, J.; Lee, T.H.; Shin, J.H.; Lee, J.W. Hot working behavior of a nitrogen-alloyed Fe–18Mn–18Cr–N austenitic stainless steel. *Mater. Sci. Eng. A* **2014**, *594*, 302–308. [CrossRef]
16. Kermanpur, A.; Behjati, P.; Han, J.; Najafizadeh, A.; Lee, Y.K. A microstructural investigation on deformation mechanisms of Fe–18Cr–12Mn–0.05 C metastable austenitic steels containing different amounts of nitrogen. *Mater. Des.* **2015**, *82*, 273–280. [CrossRef]
17. Pozuelo, M.; Wittig, J.E.; Jiménez, J.A.; Frommeyer, G. Enhanced mechanical properties of a novel high-nitrogen Cr-Mn-Ni-Si austenitic stainless steel via TWIP/TRIP effects. *Metall. Mater. Trans. A* **2009**, *40*, 1826–1834. [CrossRef]
18. Masumura, T.; Seto, Y.; Tsuchiyama, T.; Kimura, K. Work-hardening mechanism in high-nitrogen austenitic stainless steel. *Mater. Trans.* **2020**, *61*, 678–684. [CrossRef]
19. Dai, Q.; Yuan, Z.; Chen, X.; Chen, K. High-cycle fatigue behavior of high-nitrogen austenitic stainless steel. *Mater. Sci. Eng. A* **2009**, *517*, 257–260. [CrossRef]
20. Vogt, J.B. Fatigue properties of high nitrogen steels. *J. Mater. Process. Technol.* **2001**, *117*, 364–369. [CrossRef]
21. Park, J.Y.; Park, S.-J.; Kang, J.-Y.; Lee, C.-H.; Ha, H.-Y.; Moon, J.; Jang, J.H.; Lee, T.-H. Fatigue Behaviors of High Nitrogen Stainless Steels with Different Deformation Modes. *Mater. Sci. Eng. A* **2017**, *682*, 622–628. [CrossRef]
22. Hao, X.; Zhao, T.; Sun, X.; Niu, Z.; Wang, Y.; Zhang, F.; Zhao, J.; Wang, T. Adding N to Enhance Low-Cycle Fatigue Properties of High-Mn Austenitic Steel. *Mater. Sci. Eng. A* **2024**, *897*, 146357. [CrossRef]
23. Maeng, W.-Y.; Kim, M.-H. Comparative Study on the Fatigue Crack Growth Behavior of 316L and 316LN Stainless Steels: Effect of Microstructure of Cyclic Plastic Strain Zone at Crack Tip. *J. Nucl. Mater.* **2000**, *282*, 32–39. [CrossRef]
24. Zhang, W.; Li, L.; Ngai, T.; Hu, L. Influence of Hydrogen Reduction on the Properties of Porous High-Nitrogen Austenitic Stainless Steel. *Materials.* **2022**, *15*, 5585. [CrossRef]
25. Wu, X.; Fu, Y.; Huang, J.; Han, E.; Ke, W.; Yang, K.; Jiang, Z. Investigation on Pitting Corrosion of Nickel-Free and Manganese-Alloyed High-Nitrogen Stainless Steels. *J. Mater. Eng. Perform.* **2009**, *18*, 287–298. [CrossRef]
26. Olefjord, I.; Wegrelius, L. The Influence of Nitrogen on the Passivation of Stainless Steels. *Corros. Sci.* **1996**, *38*, 1203–1220. [CrossRef]
27. Yang, K.; Ren, Y. Nickel-free austenitic stainless steels for medical applications. *Sci. Technol. Adv. Mater.* **2010**, *11*, 014105. [CrossRef]

28. Chen, S.; Yao, Z.; Guan, Y.; Yang, H.; Shahzad, M.B.; Wu, Y.; Yang, K. High nitrogen stainless steel drug-eluting stent-assessment of pharmacokinetics and preclinical safety in vivo. *Bioact. Mater.* **2020**, *5*, 779–786. [CrossRef]
29. Metikoš-Huković, M.; Babić, R.; Grubač, Z.; Petrović, Ž.; Lajçi, N. High Corrosion Resistance of Austenitic Stain-less Steel Alloyed with Nitrogen in an Acid Solution. *Corros. Sci.* **2011**, *53*, 2176–2183. [CrossRef]
30. Gao, F.; Qiao, Y.; Chen, J.; Yang, L.; Zhou, H.; Zheng, Z.; Zhang, L. Effect of nitrogen content on corrosion behavior of high-nitrogen austenitic stainless steel. *npj Mater. Degrad.* **2023**, *7*, 75. [CrossRef]
31. Zhang, N.B.; Zhao, S.P.; Cai, Y.; Li, J.; Liang, X.; Lu, L.; Luo, S.N. Superior Dynamic Mechanical Properties of High Nitrogen Austenitic Stainless Steel. *Int. J. Mech. Sci.* **2025**, *307*, 110897. [CrossRef]
32. Lundberg, S.M.; Lee, S.I. A unified approach to interpreting model predictions. *arXiv* **2017**, arXiv:1705.07874. [CrossRef]
33. Ponce-Bobadilla, A.V.; Schmitt, V.; Maier, C.S.; Mensing, S.; Stodtmann, S. Practical guide to SHAP analysis: Explaining supervised machine learning model predictions in drug development. *Clin. Transl. Sci.* **2024**, *17*, e70056. [CrossRef] [PubMed]
34. Oviedo, F.; Ferres, J.L.; Buonassisi, T.; Butler, K.T. Interpretable and Explainable Machine Learning for Materials Science and Chemistry. *Acc. Mater. Res* **2022**, *3*, 597–607. [CrossRef]
35. Zhong, X.; Gallagher, B.; Liu, S.; Kailkhura, B.; Hiszpanski, A.; Han, T.Y.J. Explainable machine learning in materials science. *npj Comput. Mater.* **2022**, *8*, 204. [CrossRef]
36. Antonini, A.S.; Tanzola, J.; Asiain, L.; Ferracutti, G.R.; Castro, S.M.; Bjerg, E.A.; Ganuza, M.L. Machine Learning model interpretability using SHAP values: Application to Igneous Rock Classification task. *Appl. Comput. Geosci.* **2024**, *23*, 100178. [CrossRef]
37. Jain, A. Machine learning in materials research: Developments over the last decade and challenges for the future. *Curr. Opin. Solid State Mater. Sci.* **2024**, *33*, 101189. [CrossRef]
38. Sun, Y.; Ni, J. Machine Learning Advances in High-Entropy Alloys: A Mini-Review. *Entropy* **2024**, *26*, 1119. [CrossRef]
39. Huin, D.; Leblond, J.-B.; Darghoum, I.; Bergheau, J.-M.; Bertrand, F. Extended Wagner-Type Models and Their Application to the Prediction of the Transition from Internal to External Oxidation. *Comput. Mater. Sci.* **2022**, *209*, 111334. [CrossRef]
40. Ghassemali, E.; Conway, P.L.J. High-throughput CALPHAD: A powerful tool towards accelerated metallurgy. *Front. Mater.* **2022**, *9*, 889771. [CrossRef]
41. Liu, S.; Bocklund, B.; Diffenderfer, J.; Chaganti, S.; Kailkhura, B.; McCall, S.K.; McKeown, J.T. A comparative study of predicting high entropy alloy phase fractions with traditional machine learning and deep neural networks. *npj Comput. Mater.* **2024**, *10*, 172. [CrossRef]
42. Shen, C. The Synergy of Machine Learning and CALPHAD: Revitalizing Traditional Approaches. *Comput. Mater. Sci* **2025**, *258*, 113970. [CrossRef]
43. Sundar, A.; Tan, X.; Hu, S.; Gao, M.C. CALPHAD-based Bayesian optimization to accelerate alloy discovery for high-temperature applications. *J. Mater. Res.* **2025**, *40*, 112–122. [CrossRef]
44. Yi, W.; Ma, S.; Gao, J.; Zhong, J.; Gao, T.; Yang, S.; Li, Q. A novel atomic mobility model for alloys under pressure and its application in high pressure heat treatment Al-Si alloys by integrating CALPHAD and machine learning. *J. Mater. Sci. Technol.* **2025**, *217*, 116–127. [CrossRef]
45. Liang, Z.; Neumeier, S.; Rao, Z.; Göken, M.; Pyczak, F. CALPHAD informed design of multicomponent CoNiCr-based superalloys exhibiting large lattice misfit and high yield stress. *Mater. Sci. Eng. A* **2022**, *854*, 143798. [CrossRef]
46. Khan, F.A.; Rao, V.N.; Ghosh, A.; Karmakar, A.; Patra, S. Prediction of Mechanical Properties of Cr-Mn-N Austenitic Stainless Steel Using Machine Learning Approach. In *The International Conference on Metallurgical Engineering and Centenary Celebration*; Springer Nature: Singapore, 2023; pp. 119–130.
47. Ghadar, S.; Momeni, A.; Tolaminejad, B.; Soltanalinezhad, M. A comparative study on the hot deformation behavior of 410 stainless and K100 tool steels. *Mater. Sci. Eng. A* **2019**, *760*, 394–406. [CrossRef]
48. Wang, Y.; Zheng, H.; Cheng, L.; Wang, E.; Sheng, Z.; Wang, J.; Prahl, U. Artificial Neural Network-Based Construction of a Processing Map for P550 Steel for Non-Magnetic Drill Collars. *J. Mater. Res. Technol.* **2025**, *39*, 1543–1559. [CrossRef]
49. Jain, H.; Deb, K. An evolutionary many-objective optimization algorithm using reference-point based nondominated sorting approach, part II: Handling constraints and extending to an adaptive approach. *IEEE Trans. Evol. Comput.* **2013**, *18*, 602–622. [CrossRef]
50. Saller, G.; Aigner, H. High nitrogen alloyed steels for nonmagnetic drill collars. standard steel grades and latest developments. *Mater. Manuf. Processes.* **2004**, *19*, 41–49. [CrossRef]
51. Jargelius-Pettersson, R.F.A. Application of the pitting resistance equivalent concept to some highly alloyed austenitic stainless steels. *Corrosion* **1998**, *54*, 162–168. [CrossRef]
52. Fatima, S.; Hussain, A.; Amir, S.B.; Awan, M.G.Z.; Ahmed, S.H.; Aslam, S.M.H. XGBoost and Random Forest Algorithms: An in Depth Analysis. *Pak. J. Sci. Res.* **2023**, *3*, 26–31. [CrossRef]
53. Prokhorenkova, L.; Gusev, G.; Vorobev, A.; Dorogush, A.V.; Gulin, A. CatBoost: Unbiased boosting with categorical features. *arXiv* **2018**, arXiv:1706.09516. [CrossRef]

54. Ma, X.; Sha, J.; Wang, D.; Yu, Y.; Yang, Q.; Niu, X. Study on a prediction of P2P network loan default based on the machine learning LightGBM and XGboost algorithms according to different high dimensional data cleaning. *Electron. Commer. Res. Appl.* **2018**, *31*, 24–39. [CrossRef]
55. Liu, Z.; Tian, W.; Cui, Z.; Liu, B. A universal microkinetic-machine learning bimetallic catalyst screening method for steam methane reforming. *Sep. Purif. Technol.* **2023**, *311*, 123270. [CrossRef]
56. Peng, Y.; Zhou, W.; Cao, X.; Liu, M.; Zhang, H.; Zhang, Y.; Liu, X. Machine learning for predicting separation factors of chiral diphosphine ligands in chiral extraction of amino acid and mandelic acid enantiomers. *Sep. Purif. Technol.* **2025**, *355*, 129797. [CrossRef]
57. Cai, Q.Y.; Qiao, L.Z.; Yao, S.J.; Lin, D.Q. Machine learning assisted QSAR analysis to predict protein adsorption capacities on mixed-mode resins. *Sep. Purif. Technol.* **2024**, *340*, 126762. [CrossRef]
58. Nohara, Y.; Matsumoto, K.; Soejima, H.; Nakashima, N. Explanation of machine learning models using shapley additive explanation and application for real data in hospital. *Comput. Methods Programs Biomed.* **2022**, *214*, 106584. [CrossRef]
59. Liu, C.; Wang, X.; Cai, W.; Yang, J.; Su, H. Optimal design of the austenitic stainless-steel composition based on machine learning and genetic algorithm. *Materials* **2023**, *16*, 5633. [CrossRef]
60. Deb, K.; Jain, H. An evolutionary many-objective optimization algorithm using reference-point-based nondominated sorting approach, part I: Solving problems with box constraints. *IEEE Trans. Evol. Comput.* **2013**, *18*, 577–601. [CrossRef]

*materials*

MDPI

*Article*

# Effect of Si on Marine Corrosion Behavior of Austenite Low-Density Steel

**Yuhe Huang, Shuize Wang, Jiahao Qiang, Hui Wang and Jun Lu ***

Institute for Carbon Neutrality, University of Science and Technology Beijing, Beijing 100083, China; huangyuhe@ustb.edu.cn (Y.H.); wangshuize@ustb.edu.cn (S.W.); 18500713877@139.com (J.Q.); wanghui04@ustb.edu.cn (H.W.)
* Correspondence: lujun@ustb.edu.cn; Tel.: +86-17801024638

**Abstract**

To address the high-salinity and hyper-humid thermal environment of tropical oceans and meet industrial demands for high strength and lightweight, austenitic low-density steel was developed as a novel corrosion-resistant steel. A 3.5 wt.% NaCl solution was used to simulate the marine environment to study the effect of Si on the corrosion behavior of this steel. Scanning electron microscopy (SEM), transmission electron microscopy (TEM), X-ray photoelectron spectroscopy (XPS), X-ray diffraction (XRD) and electron probe microanalysis (EPMA) were employed to characterize the microstructures and corrosion behaviors of two test steels, as well as the phase compositions and element distributions of corrosion products after polarization and cyclic immersion accelerated corrosion tests. The results show that a dense oxide film initially forms on the steel surface in 3.5 wt.% NaCl solution at the early corrosion stage. Si addition induces $SiO_2$ formation and promotes Al conversion to $Al_2O_3$, enhancing oxide film compactness and inhibiting matrix atom outward diffusion and $Cl^-$ inward penetration. With prolonged corrosion, the oxide film is dissolved or broken, forming a dense rust layer dominated by $Fe_3O_4$, $Fe_2O_3$ and FeOOH. Si enriches in the inner rust layer adjacent to the matrix and pitting cavities, inhibiting pitting deepening and promoting $\gamma$-FeOOH to $\alpha$-FeOOH transformation, thus improving the steel's corrosion resistance.

**Keywords:** austenite low-density steel; marine corrosion; surface oxidation; phase composition; element enrichment

## 1. Introduction

Marine platforms, as essential infrastructure for offshore oil and gas exploitation, offshore wind power development, and other marine engineering activities, play an irreplaceable role in energy development, resource transportation, environmental monitoring, and the safeguarding of national maritime defense security [1,2]. According to their connection with the seabed, marine platforms can be classified into fixed platforms and floating platforms, which are, respectively, suited to different marine environments and operational requirements [3–6]. As marine engineering advances toward deeper and more remote offshore environments with increasingly complex service conditions, the performance requirements for steels used in marine platforms continue to rise, demanding higher strength, longer service life, and superior environmental resistance [7–10]. Harsh marine corrosion environments characterized by high salinity, high humidity, and elevated temperatures create highly corrosive conditions. Under such circumstances, the synergistic effects of multiple corrosion factors can accelerate the corrosion-induced degradation of platform steels, thereby compromising platform safety and service life and increasing costs and safety risks. Consequently, more stringent requirements are imposed on the corrosion resistance of steels used for marine platforms [11,12].

Lightweight design, long service life, and high specific strength are the core developmental requirements for steels used in marine platforms at present [13,14]. Marine platform structures are massive in scale, and the self-weight of the equipment directly affects load-bearing capacity, construction difficulty, and operational costs. Reducing material density and increasing specific strength can decrease structural weight while maintaining high strength and reliability, thereby enabling lightweight platform design. Meanwhile, a long service life can reduce the frequency of maintenance and replacement caused by corrosion-induced material failure, lower operation and maintenance costs, and ensure the long-term stable service of marine equipment. Austenitic low-density steel, as a novel high-performance structural steel, is well aligned with the above development requirements. Its key advantages stem from its relatively low density, high specific strength, and excellent corrosion resistance. These attributes help overcome major limitations of conventional marine steels, including high density, low specific strength, and poor corrosion resistance in harsh marine environments. As a result, it is a promising material for lightweight, long-service-life marine platforms.

Low-density Fe–Mn–Al–C austenitic steels, developed as low-cost alternatives to Fe–Cr–Ni–C stainless steels since the 1980s and 1990s, have attracted considerable attention, primarily owing to the pronounced beneficial effects of Mn and Al on the mechanical properties and oxidation resistance of these materials. Among these elements, Al exhibits a particularly significant weight-reduction effect: for every 1 wt.% increase in Al content, the material density can be reduced by approximately 1.3% [15]. In addition to Al, Si is also an important alloying element for reducing material density. Si provides solid-solution strengthening, thereby enhancing material strength, while simultaneously increasing the activity of carbon in austenite, hindering the nucleation and growth of carbides, suppressing carbide formation, and promoting carbon enrichment in untransformed austenite, which significantly improves the stability of retained austenite [16]. Moreover, Si possesses high oxide stability and can form a dense $SiO_2$ oxide film during high-temperature oxidation, effectively reducing the corrosion rate of the material; it can also shift the steady-state potential in the negative direction, facilitating the transition of metallic materials into a stable passive state [17]. In addition, the presence of Si in steel, together with fine grains, can promote the continuity of the $Al_2O_3$ oxide film, whereas a discontinuous alumina layer does not provide effective corrosion protection [18]. Furthermore, during oxidation, Si oxidizes at a faster rate than Al, enabling the rapid formation of an oxide film.

At present, studies on the role of Si in low-density Fe–Mn–Al–C steels mainly focus on its contributions to mechanical properties and high-temperature oxidation behavior, while systematic investigations into the effects of Si on long-term corrosion behavior and the evolution mechanisms of microscopic passive films under simulated harsh marine environments remain lacking. These limitations result in an insufficiently close linkage between alloy composition design and actual severe service conditions, thereby restricting the reliable application of this class of steels in marine platforms. In chloride-rich marine environments (e.g., seawater immersion or marine atmospheres), the corrosion response of Fe–Mn–Al–(Si) low-density steels is governed by the stability and chemistry of the surface film and corrosion products, and is often evaluated in 3.5 wt.% NaCl as a laboratory surrogate [19,20]. Although widely adopted to stabilize austenite in many Fe–Mn–Al–C lightweight alloys, Mn can lower corrosion resistance because Mn tends to form less protective/unstable oxides and promotes more active dissolution in chloride media. By contrast,

https://doi.org/10.3390/ma19061182

Al can enhance corrosion protection when it enriches the passive film and contributes to compact $Al_2O_3$-based layers (or Al-enriched oxides), which strengthens passivation and delays chloride-induced breakdown in Fe–Mn–Al–C lightweight alloys [21]. Silicon typically improves resistance to chloride attack mainly as a film/rust modifier: it can refine the passive layer and raise the pitting potential (expanding the passive region), while also promoting $SiO_2$-containing network structures and altering ionic selectivity and elemental partitioning within the oxide/rust layer [22]. Recent composition-design work on Fe–Mn–Al–Cr–Si–Mo–C lightweight steels further indicates that coupling Mn-based austenite stabilization with Al/Si-assisted film stabilization can significantly shift polarization behavior and corrosion mechanisms in 3.5 wt.% NaCl, underscoring the need for balanced alloy design for marine service [23].

In this study, austenitic low-density steel was selected as the research object, and a 3.5 wt.% NaCl solution was used to simulate the marine environment. The corrosion resistance of the material was evaluated through electrochemical experiments, in which potentiostatic polarization tests were employed to simulate the initial stage of corrosion under marine conditions, while a periodic immersion accelerated corrosion test was conducted to simulate long-term corrosion behavior in a marine environment. The effects of Si addition on the oxide film compactness, elemental distribution, phase constitution of corrosion products, and surface morphology of derusted steel in a marine environment were systematically investigated. The influence of Si addition on the corrosion behavior of austenitic low-density steel in a marine environment was elucidated, and the corrosion mechanism of Si-containing low-density steel under marine conditions was clarified, providing a theoretical basis and technical guidance for engineering applications in related fields.

## 2. Materials and Methods

### 2.1. Materials

Two austenitic low-density steels were designed in the laboratory, one without Si and the other containing 0.8 wt.% Si; the chemical compositions of the experimental steels measured by chemical method are listed in Table 1. The steel was produced by vacuum smelting and casting (VIM-25, Beijing Research Institute of Mechanical and Electrical Technology Co., Ltd., Beijing, China). Prior to forging, the steels were homogenized at 1150–1200 °C for 12 h to prevent Mn and Al segregation. After homogenization, forging was carried out, and the forged billets were reheated in a furnace to 1150 °C, held for 2 h, descaled, and then subjected to rolling. Following five passes of hot rolling, with a final rolling temperature of no less than 950 °C, the thickness was reduced from 60 mm to 12 mm, after which the plates were water-cooled to room temperature. The density of the two test steels were measured by the Archimedes method as: 0Si steel was 6.83 $g/cm^3$, while that of the 0.8Si steel was 6.80 $g/cm^3$. For EBSD characterization, the specimens were prepared sequentially by mechanical grinding (using SiC papers from 250# to 2000#), mechanical polishing, and vibratory polishing (VibroMet 2, Buehler, Lake Bluff, IL, USA) with 50 nm $SiO_2$ particles. Tensile specimens were prepared in accordance with GB/T 228.1-2021 [24] (specimen diameter: 10 mm; gauge length: 50 mm). Impact specimens were prepared as standard Charpy specimens in accordance with GB/T 229-2020 [25].

**Table 1.** Chemical composition of experimental steel (wt.%).

| wt.% | C | Si | Mn | Al | Fe |
|---|---|---|---|---|---|
| 0Si | 0.9 | 0 | 29.0 | 7.5 | Bal. |
| 0.8Si | 0.9 | 0.8 | 29.0 | 7.5 | Bal. |

https://doi.org/10.3390/ma19061182

### 2.2. Electrochemical Experiments

Electrochemical experiments were conducted to evaluate the intrinsic corrosion resistance of the experimental steels. The specimens for electrochemical testing were prepared solely by mechanical grinding up to 2000# grit. Electrochemical corrosion tests were carried out in a 3.5 wt.% NaCl solution using an Reference 600+ electrochemical workstation (Gamry Instruments, Warminster, PA, USA). A traditional three-electrode system was used for electrochemical measurements, in which a saturated calomel electrode (SCE) was used as reference electrode, a platinum plat was used as counter electrode, and the test steels were used as working electrode, the working electrode area was 0.785 $cm^2$. Prior to testing, the specimens were immersed for 30 min to allow the electrode potential to stabilize, which was taken as the open-circuit potential (OCP). Electrochemical impedance spectroscopy (EIS) measurements were performed over a frequency range from $10^5$ Hz to 0.01 Hz, with an AC perturbation amplitude of $\pm$10 mV, under open-circuit potential conditions. Potentiodynamic polarization (PDP) curves were recorded in the potential range of $-1.5$ to 0.5 V at a scan rate of 1 mV/s. Each sample is tested three times to ensure data reliability. All electrochemical tests were carried out at a constant temperature of 25 $\pm$ 1 °C, with the ambient relative humidity controlled at 45–55% RH.

### 2.3. Potentiostatic Polarization Tests

Oxide films were prepared by potentiostatic polarization tests. Prior to the experiments, the specimens were ground to 3000# grit using abrasive papers and thoroughly cleaned with deionized water to remove surface debris. The tests were relative to the OCP, before potentiostatic polarization, the samples were cathodically polarized at $-1.0$ $V_{SCE}$ for 300 s to remove the air-formed oxide film, followed by polarization at $-0.5$ $V_{SCE}$ for 7200 s to form an oxide film on the surface. After polarization, the samples were immediately rinsed with deionized water, dried with absolute ethanol, and then vacuum-sealed to prevent surface oxidation. Surface morphology characterization and compositional analysis were performed using a scanning electron microscope (FE-SEM, Mira 3 LMS, TESCAN, Brno, Czech Republic) equipped with energy-dispersive X-ray spectroscopy (EDS, Ultim Max 170, Oxford Instruments, Abingdon, UK). The phase composition of the oxide film surface was analyzed by X-ray photoelectron spectroscopy (XPS, Thermo Scientific ESCALAB 250Xi, Thermo Fisher Scientific, Waltham, MA, USA); argon ion sputtering was employed, with spectra collected after etching every 6 nm in depth, to a total etching depth of 30 nm, in order to examine the variation in elemental distribution with oxide film depth. Transmission specimens were prepared by focused ion beam (FIB, Helios 5 UC, Thermo Fisher Scientific, Waltham, MA, USA) milling, and the cross-sectional morphology and compositional features of the oxide films were observed using a transmission electron microscope equipped with EDS (FE-TEM, Talos F200X, Thermo Fisher Scientific, Waltham, MA, USA).

### 2.4. Periodic Immersion Accelerated Corrosion Test

The experimental steels were machined into two specimen sizes: large specimens with dimensions of 50 mm $\times$ 25 mm $\times$ 3 mm (length $\times$ width $\times$ thickness) and small specimens with dimensions of 11 mm $\times$ 10 mm $\times$ 3 mm (length $\times$ width $\times$ thickness). A circular hole with a diameter of 2.5 mm was machined at the center of the width side of each specimen for suspension during testing. For each experimental condition and each corrosion cycle, three parallel large specimens were prepared to calculate corrosion weight loss, and the scraped rust powders were used for phase analysis; meanwhile, three parallel small specimens were prepared for morphological observation and elemental distribution analysis. Prior to the periodic immersion test, all specimens were surface-machined and

https://doi.org/10.3390/ma19061182

ultrasonically cleaned with ethanol and deionized water to remove surface contaminants. A 3.5 wt.% NaCl solution was selected to simulate the marine environment. To further accelerate the corrosion process, the corrosion medium temperature was set to 45 °C, the drying temperature to 70 °C, and the relative humidity to 70%, in order to simulate a more severe high-humidity and high-temperature marine corrosion environment. Before the periodic immersion corrosion test, all large specimens were weighed using a precision electronic balance to record the initial mass for subsequent weight-loss measurements. After the completion of the corrosion tests, the specimens were removed and the surface rust layers were eliminated using a chemical derusting method; the derusting solution consisted of 500 mL HCl + 3.5 g $C_6H_{12}N_4$ (hexamethylenetetramine) + 500 mL $H_2O$. The derusting process was conducted at room temperature (25 °C) with ultrasonic agitation to facilitate the removal of surface-adhered corrosion products. The corrosion durations were 72, 144, 216, 288, and 360 h. The macroscopic morphologies of the corroded specimens were recorded using a digital camera (EOS 90D, Canon Inc., Tokyo, Japan). The phase composition and chemical composition of the rust layers were analyzed by X-ray diffraction (XRD, X'Pert PRO MPD, Malvern Panalytical, Almelo, The Netherlands) and X-ray photoelectron spectroscopy (XPS, Thermo Scientific ESCALAB 250Xi, Thermo Fisher Scientific, Waltham, MA, USA), respectively. The cross-sectional morphology of the rust layers and their elemental distributions were characterized using an electron probe microanalyzer (EPMA, JXA-8530F Plus, JEOL Ltd., Tokyo, Japan). The surface morphology of the derusted specimens was observed using a three-dimensional laser confocal microscope (OLS5100, Olympus Corporation, Tokyo, Japan), and the pit depths were statistically analyzed.

## 3. Results

### 3.1. Microstructure and Mechanical Properties

Figure 1 presents the microstructural morphologies and tensile curves of the two experimental steels. The results indicate that the addition of 0.8 wt.% Si has almost no effect on the microstructural morphology of the austenitic low-density steel; both hot-rolled low-density steels exhibit a single-phase austenitic microstructure. The austenite grain size is inhomogeneous, consisting of large deformed grains formed by thermal deformation during the rolling process, as well as fragmented fine grains. As a solid-solution element, Si enhances the material strength through solid-solution strengthening. It suppresses carbide precipitation and promotes carbon enrichment and stabilization of austenite [26]. Consequently, the 0.8Si steel shows a slight increase in ultimate tensile strength, accompanied by a slight decrease in elongation after fracture and total elongation. The mechanical properties of the two experimental steels are summarized in Table 2, including Yield strength ($Rp_{0.2}$), tensile strength ($R_m$), elongation after fracture ($A_{50}$), uniform elongation ($A_g$), impact energy at 0 °C ($KV_2$, 0 °C), specific strength ($Rm/\rho$). The 0Si steel exhibits an ultimate tensile strength of 1010 MPa, an elongation after fracture of 51.9%, and an impact energy at 0 °C of 148.2 J. In comparison, the 0.8Si steel shows an ultimate tensile strength of 1029 MPa, an elongation after fracture of 50.5%, and an impact energy at 0 °C of 129.2 J, which is slightly lower than that of the 0Si steel. The addition of Si reduces the density of the experimental steel, and the 0.8Si steel exhibits a higher ultimate tensile strength; therefore, its specific strength (151.3 N·m/kg) is higher than that of the 0Si steel (147.9 N·m/kg).

https://doi.org/10.3390/ma19061182

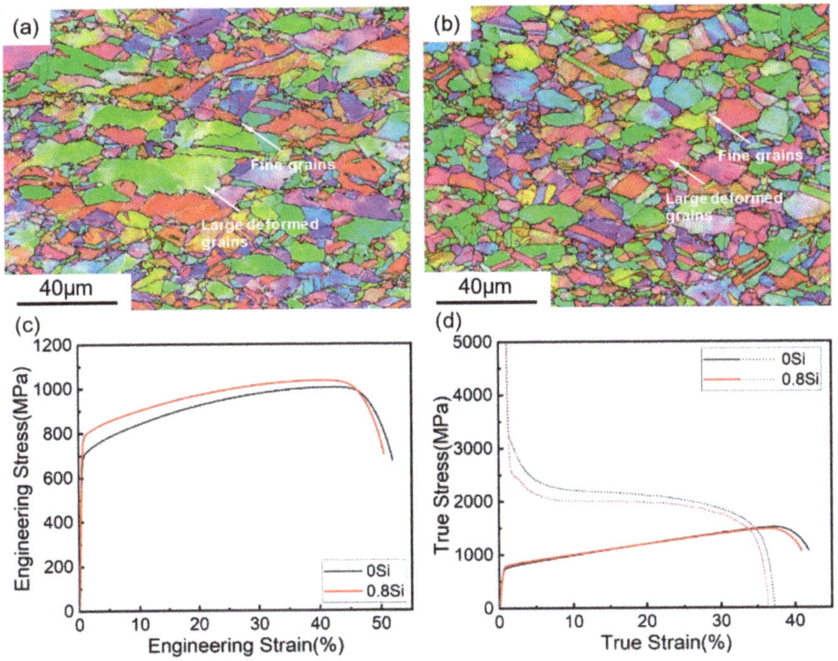


**Figure 1.** Microstructure and stress curve of experimental steel. (**a**) 0Si; (**b**) 0.8Si; (**c**) stress–strain curves; (**d**) work-hardening curves.

**Table 2.** Mechanical properties of experimental steel.

| Steel | $Rp_{0.2}$ (MPa) | $R_m$ (MPa) | $A_{50}$ (%) | $A_g$ (%) | $KV_2$ (0 °C) (J) | Rm/ρ (N·m/kg) |
|---|---|---|---|---|---|---|
| 0Si | 697 | 1010 | 51.9 | 41.4 | 148.2 | 147.9 |
| 0.8Si | 751 | 1029 | 50.5 | 40.4 | 129.2 | 151.3 |

### 3.2. Electrochemical Experimental Results

The potentiodynamic polarization curves and electrochemical impedance spectroscopy (EIS) results (Nyquist and Bode plots) of the two experimental steels are shown in Figure 2. From the potentiodynamic polarization curves, the corrosion current density (icorr), corrosion potential (Ecorr), and pitting potential (Ep) of the two steels were determined, and the results are summarized in Table 3. As indicated by Figure 2a and Table 3, after the addition of Si, both Ecorr and Ep increase, while icorr and ip decreases, and a distinct passive region is observed for the experimental steels [27]. The passive current density of the 0.8Si steel is lower than that of the 0Si steel, indicating that the addition of Si to austenitic low-density steel affects both the intrinsic corrosion behavior and the characteristics of the passive film.

https://doi.org/10.3390/ma19061182

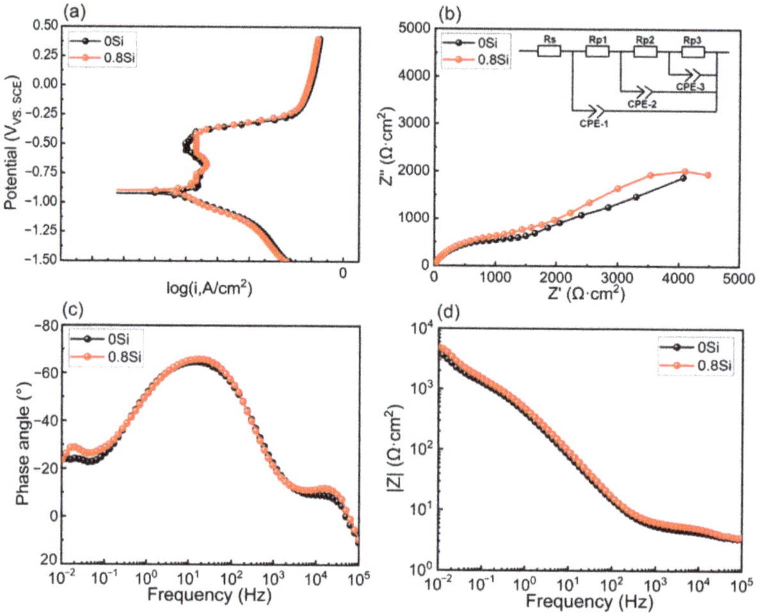


**Figure 2.** Electrochemical results for the experimental steel. (**a**) Potentiodynamic polarization curves; (**b**) impedance spectra and equivalent circuit; (**c**) phase angle plots; (**d**) impedance modulus plots.

**Table 3.** The electrochemical parameters of experimental steel.

| Steel | Ecorr ($V_{vs.SCE}$) | Icorr (μA/cm$^2$) | Ep/ ($V_{vs.SCE}$) | Ip/ (μA/cm$^2$) |
|---|---|---|---|---|
| 0Si | $-0.93 \pm 0.024$ | $4.28 \pm 1.233$ | $-0.48 \pm 0.018$ | $27.79 \pm 2.563$ |
| 0.8Si | $-0.90 \pm 0.020$ | $3.84 \pm 0.985$ | $-0.38 \pm 0.016$ | $20.94 \pm 3.070$ |

As shown in Figure 2b,c, the Nyquist and phase angle plots of the 0.8Si steel exhibit three arc segments, whereas the 0Si steel shows a straight line in the low-frequency region and arc-shaped features in the middle- and high-frequency regions. The arc in the high-frequency region corresponds to the charge-transfer process between the working electrode surface and the solution. Because an oxide film rapidly forms on the steel surface, the diameter of the high-frequency arc is extremely small and difficult to discern in the Nyquist plot; however, it can be clearly observed in the phase angle plot. The diameter of the high-frequency arc for the 0.8Si steel is larger than that for the 0Si steel, indicating a slower electron transfer rate between the electrode surface and the solution, which ultimately leads to the formation of a more uniform oxide film [28]. The arc in the medium-frequency region corresponds to the film impedance of the surface oxide layer [29–31], and the 0.8Si steel exhibits a larger arc radius, indicating a higher oxide film resistance. The low-frequency arc corresponds to the impedance of the corrosion product deposition layer formed after the gradual dissolution of the oxide film. The appearance of a low-frequency arc for the 0.8Si steel, in contrast to a straight line for the 0Si steel, indicates that the surface reaction at this stage is extremely rapid and no longer controlled by charge transfer but by ionic diffusion. This behavior demonstrates that the corrosion product deposition layer on the surface of the 0Si steel contains a higher density of pores, allowing a large number of solution ions to penetrate into the pores and react with the substrate [32].

### 3.3. *Oxide Films Prepared by Potentiostatic Polarization*

#### 3.3.1. Observation of Surface Oxide Film Morphology

According to the electrochemical results, austenitic low-density steel initially forms a dense oxide film; subsequently, the oxide film dissolves or is locally broken down, followed by the formation of external corrosion products at later stages, ultimately leading to the development of a rust layer. Potentiostatic polarization tests were employed to evaluate the initial corrosion behavior of austenitic low-density steel and to prepare oxide films formed at the early stage of corrosion. Figure 3 shows the relationship between corrosion current density and time. Within the first 200 s of polarization, the current density of the experimental steels rapidly decreases to approximately 3 $\mu A/cm^2$, corresponding to the formation of a surface oxide film and the initial anodic dissolution. This phenomenon indicates that austenitic low-density steel can rapidly form an oxide film within a short period, thereby suppressing anodic dissolution. With increasing polarization time, the current density gradually increases, indicating the progressive formation of corrosion products outside the oxide layer. The slight fluctuations observed in the polarization curves correspond to the breakdown and self-healing processes of the oxide film. Ultimately, the 0.8Si steel exhibits a lower steady-state current density, reflecting the higher compactness and relatively stronger protective capability of its oxide film.

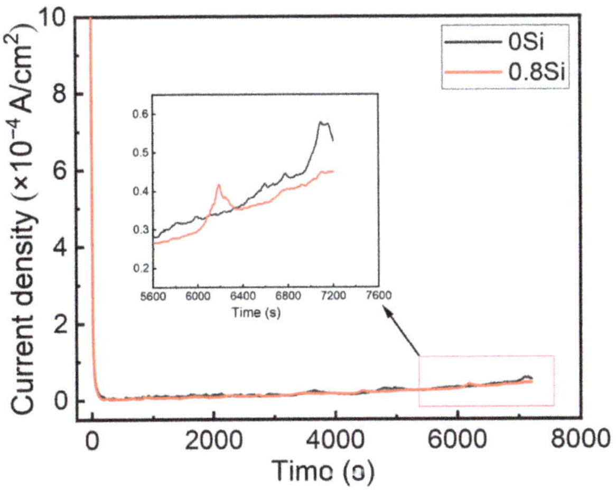


**Figure 3.** Potentiostatic polarization curves of experimental steel in 3.5 wt.% NaCl solution.

Cross-sectional transmission electron microscopy analyses were performed on the oxide films formed on the surfaces of the two experimental steels after polarization, and high-angle annular dark-field (HAADF) imaging together with scanning transmission electron microscopy energy-dispersive X-ray spectroscopy (STEM–EDS) mapping were obtained, as shown in Figure 4. The STEM–EDS results confirm the presence of oxygen in the films and reveal a uniform distribution of all alloying elements within the oxide layer [33]. From the morphological images, it can be clearly observed that the oxide film on the 0Si steel exhibits pronounced thickness inhomogeneity, with a large variation in thickness, ranging from approximately 75 nm at the thinnest region to about 127 nm at the thickest region. Such a highly nonuniform cross-sectional morphology directly reflects the presence of numerous structural defects in the oxide film on this steel, indicating relatively poor compactness of the film. The thinner regions of the oxide film represent weak points in resisting the attack of corrosive media; in $Cl^-$ containing environments, these vulnerable sites are readily penetrated and disrupted by $Cl^-$ ions, leading to the forma-

https://doi.org/10.3390/ma19061182

tion of pitting sites on the steel substrate and serving as initiation points for subsequent corrosion processes. In contrast, the oxide film formed on the surface of the 0.8Si steel exhibits not only a significantly increased thickness, reaching approximately 138 nm, but also a much more uniform thickness distribution, overall presenting a flat and compact cross-sectional morphology. This observation clearly demonstrates that the addition of Si to low-density steel effectively optimizes the growth process of the oxide film, promoting the formation of a protective layer with fewer defects and a more compact and stable structure. Such a dense and uniform oxide film can more effectively hinder the ingress of aggressive ions such as $Cl^-$, substantially reducing the probability of ion penetration through the film and thereby exhibiting superior resistance to pitting corrosion during subsequent corrosion exposure.

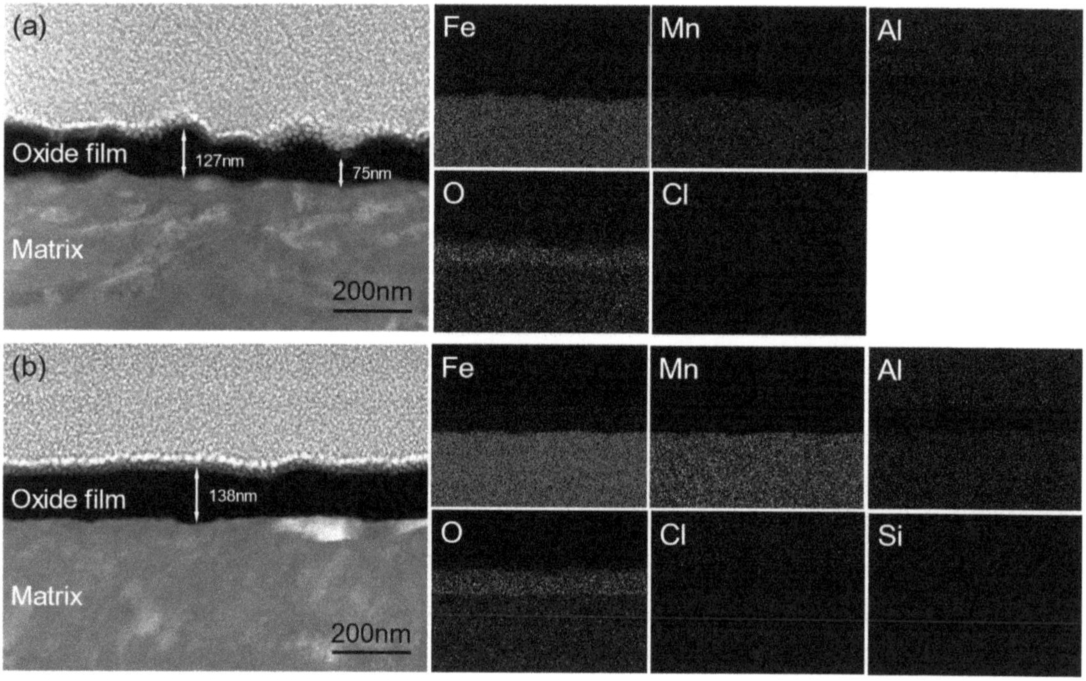


**Figure 4.** Morphology and elemental distribution of the experimental steel after polarization. (**a**) 0Si; (**b**) 0.8Si.

### 3.3.2. Phase Analysis of Surface Oxide Films

To further determine the specific phase constituents of the oxide films and the relative contents of the constituent elements, X-ray photoelectron spectroscopy (XPS) analyses were conducted on the surface oxide films formed on the two experimental steels after polarization. Figure 5 presents the XPS spectra of Fe $2p_{3/2}$, Mn $2p_{3/2}$, Al 2p, and O 1s for the oxide film on the 0Si steel. According to the Fe $2p_{3/2}$ spectrum of the oxide film on the 0Si steel (Figure 5a), Fe is present in the forms of FeO (707.0 eV), $Fe_2O_3$ (710.2 eV), FeOOH (712.7 eV), and a satellite peak (715.9 eV) [34]. These results indicate that Fe has undergone further corrosion; not only has an oxide film formed on the steel surface, but a considerable amount of Fe has also diffused outward and been further corroded outside the oxide film, suggesting that the oxide film formed on the 0Si steel is insufficiently compact to effectively inhibit the outward diffusion of substrate elements. Figure 5b shows the Mn $2p_{3/2}$ XPS spectrum of the oxide film on the 0Si steel, which is mainly composed of

https://doi.org/10.3390/ma19061182

$MnO_2$ (641.8 eV), with the satellite peak at 646.2 eV corresponding to $Mn(OH)_2$. The Al 2p spectrum (Figure 5c) exhibits a single peak attributed to $Al_2O_3$ (74.3 eV). According to the O 1s spectrum (Figure 5d), oxygen in the surface oxide film exists in the forms of $O^{2-}$ (530.6 eV) and $OH^-$ (531.6 eV), corresponding to oxides and hydroxides, respectively.

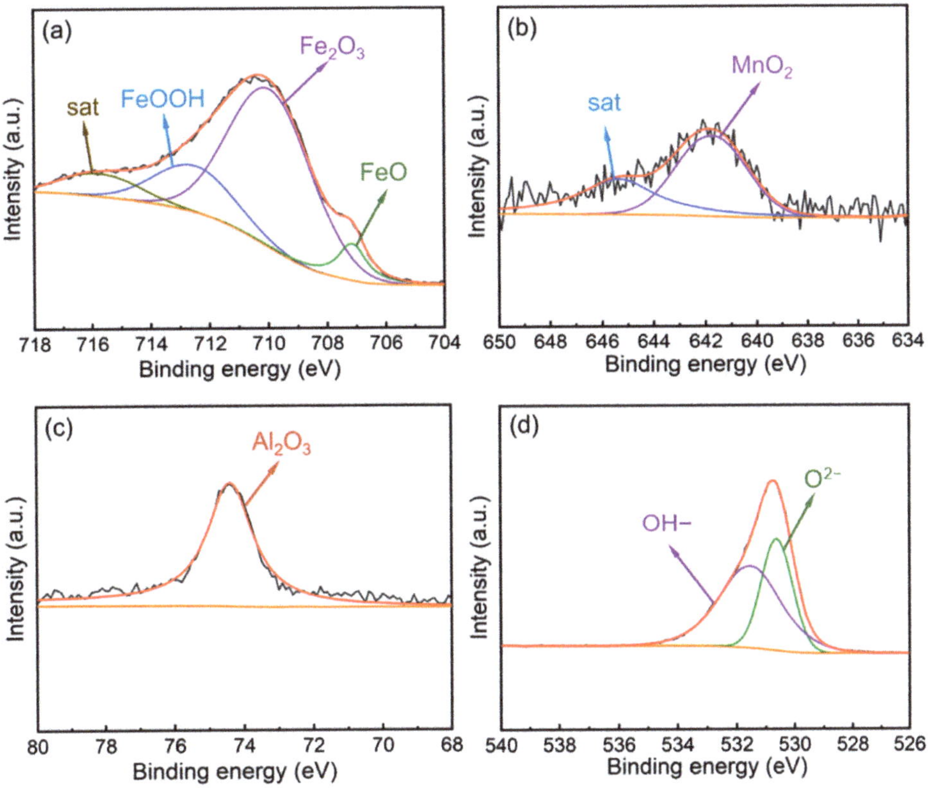


**Figure 5.** XPS spectra of the oxide film of 0Si experimental steel. (**a**) Fe $2p_{3/2}$; (**b**) Mn $2p_{3/2}$; (**c**) Al 2p; (**d**) O 1s.

According to the Fe $2p_{3/2}$ spectrum of the oxide film on the 0.8Si steel (Figure 6a), Fe is present only in the forms of FeO (707.0 eV) and $Fe_2O_3$ (710.2 eV), with no FeOOH peak detected, indicating that only an oxide film is formed on the steel surface with good compactness, which effectively prevents the outward diffusion of substrate elements. Figure 6b shows the Mn $2p_{3/2}$ XPS spectrum of the oxide film on the 0.8Si steel, which is mainly composed of MnO (640.2 eV) and satellite peaks. The Al 2p spectrum (Figure 6c) also exhibits a single peak corresponding to $Al_2O_3$ (74.3 eV). The Si 2p spectrum (Figure 6d) presents a single peak attributed to $SiO_2$ (103.4 eV), demonstrating that the Si added to austenitic low-density steel participates in the formation of the oxide film at the initial stage of corrosion. The formed $SiO_2$ is amorphous; compared with the crystalline structures of other oxides, the amorphous phase can more effectively bond with the substrate and surrounding crystalline phases, thereby enhancing the compactness and stability of the oxide film. The O 1s spectrum (Figure 6e) consists of peaks corresponding to $O^{2-}$ (530.6 eV), $OH^-$ (531.6 eV), and $H_2O$ (533.7 eV), which are associated with oxides, hydroxides, and crystallization water, respectively.

https://doi.org/10.3390/ma19061182

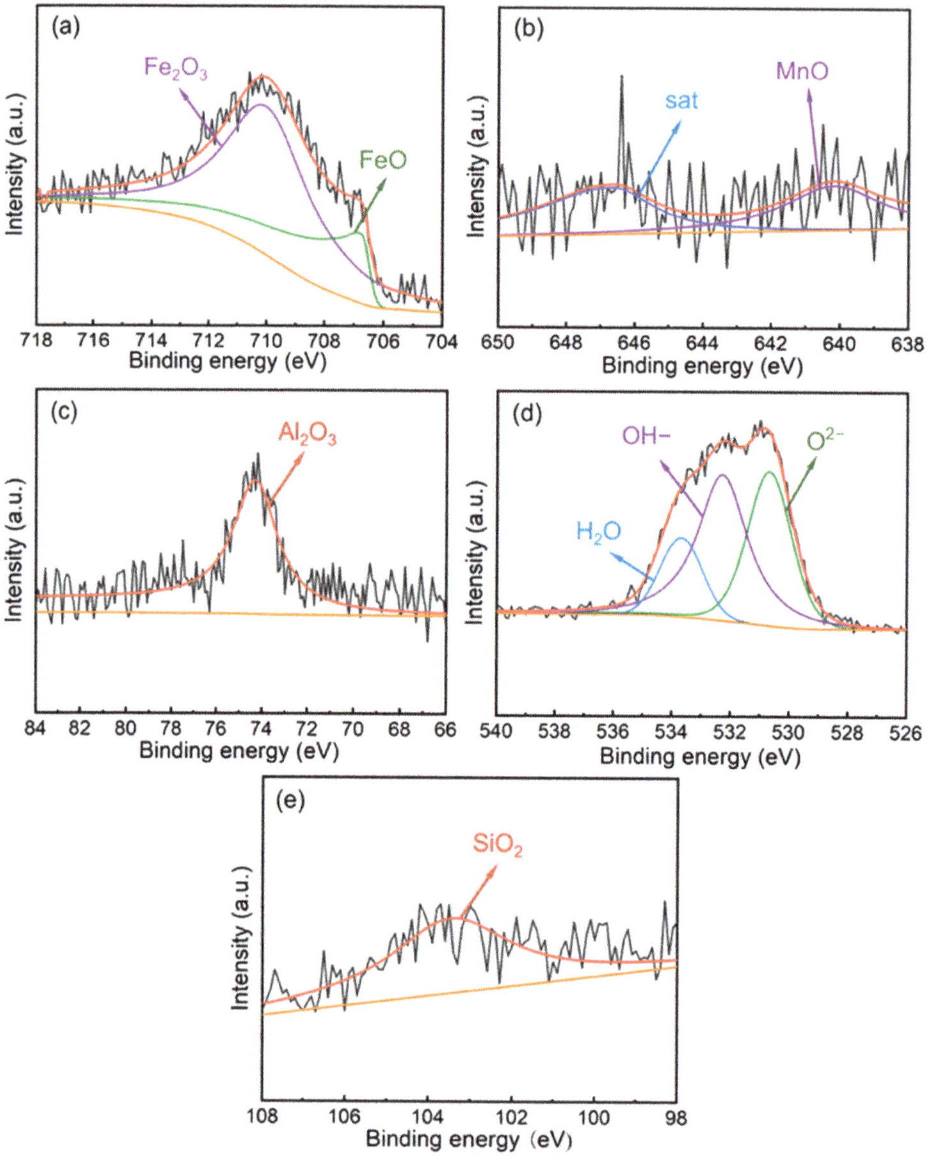


**Figure 6.** XPS spectra of the oxide film of 0.8Si experimental steel. (**a**) Fe $2p_{3/2}$; (**b**) Mn $2p_{3/2}$; (**c**) Al 2p; (**d**) O 1s (**e**) Si 2p.

### *3.4. Results of the Periodic Immersion Accelerated Corrosion Test*

#### 3.4.1. Measurement of Corrosion Weight Loss and Calculation of Corrosion Rate

After the periodic immersion accelerated corrosion tests in a 3.5 wt.% NaCl solution, the mass loss and corrosion rate variations in the two experimental steels at different corrosion durations are shown in Figure 7. As illustrated in Figure 7a, the mass loss of the experimental steels increases progressively with increasing corrosion duration, and at each corrosion interval, the mass loss of the 0.8Si steel is significantly lower than that of

https://doi.org/10.3390/ma19061182

the 0Si steel. Based on the mass loss data of the experimental steels at different corrosion durations, the corrosion rates were calculated using Equation (1).

$$v = \frac{\Delta m}{t \times S} \quad (1)$$

where $v$ is the corrosion rate ($g \cdot m^{-2} \cdot h^{-1}$); $\Delta m$ is the mass loss of the experimental steel before and after corrosion (g); $t$ is the corrosion time (h); and $S$ is the surface area of the experimental steel ($m^2$). Figure 7b shows the variation in corrosion rate of the experimental steels at different corrosion durations, exhibiting a trend of an initial decrease followed by an increase. This behavior can be attributed to the rapid formation of a dense oxide film on the steel surface at the early stage of corrosion, which effectively inhibits the outward diffusion of metal atoms from the substrate. The oxide film formed on the 0.8Si steel provides superior protection compared with that on the 0Si steel; therefore, during the period from 72 to 216 h, the corrosion rates of both steels decrease, with a more pronounced decline observed for the 0.8Si steel. When corrosion proceeds for a sufficiently long duration, rust layer spallation occurs, and the oxide film has largely dissolved. Consequently, aggressive $Cl^-$ ions from the environment gradually penetrate the surface, and metal elements from the substrate diffuse outward and react with oxygen, leading to an increase in the corrosion rate.

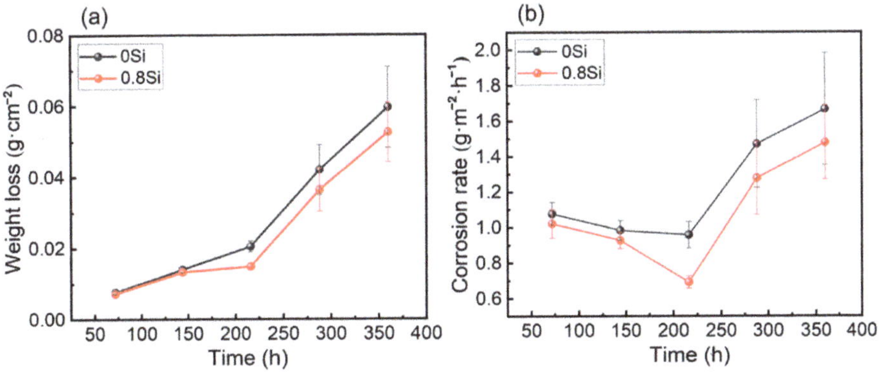


**Figure 7.** Mass loss and corrosion rate of experimental steel in different exposure time. (**a**) Weight loss; (**b**) corrosion rate.

### 3.4.2. Morphology and Elemental Distribution of Corrosion Products

Figure 8 shows the macroscopic morphologies of the two experimental steels after periodic immersion accelerated corrosion tests at different corrosion durations. After 72 h of periodic immersion, rust layers formed on the specimen surfaces, which were mainly yellow–brown in color, and corrosion pits were clearly observed; the number of corrosion pits on the 0.8Si steel was significantly lower than that on the 0Si steel. As the corrosion duration increased to 144 h, part of the rust layer began to transform into a brown–black color. After 216 h of exposure in the saline environment, most of the rust layer had transformed into brown–black rust; however, partial rust spallation occurred in the smooth, pit-free regions on the surface of the 0Si steel, exposing the substrate, whereas such spallation was not evident on the 0.8Si steel. Due to rust layer spallation and exposure of the substrate, corrosion was re-initiated at these locations; therefore, after 288 h, the steel surfaces exhibited newly formed yellow–brown rust layers, corresponding to corrosion products generated over a relatively short exposure time. After 360 h of exposure in the

https://doi.org/10.3390/ma19061182

saline environment, owing to the prolonged exposure duration, almost the entire surface was transformed into brown–black rust layers.

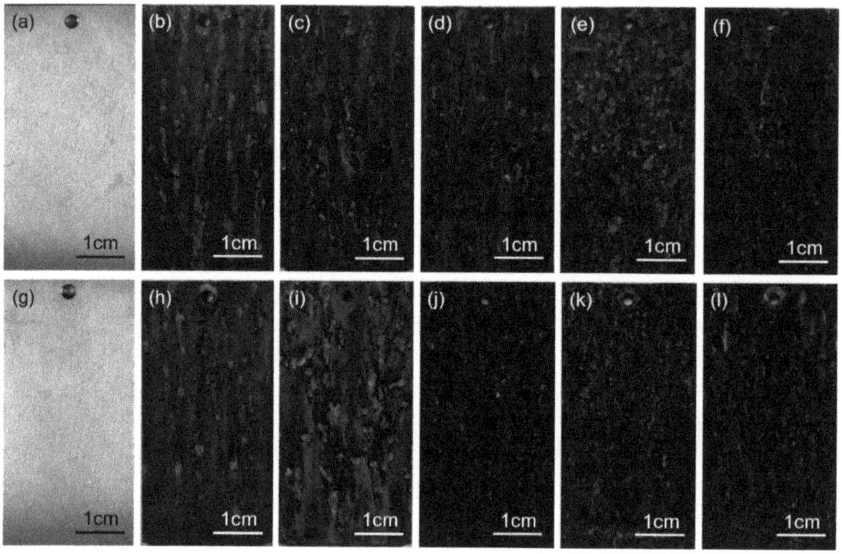


**Figure 8.** Macroscopic morphology of experimental steel in different corrosion times. (**a**) 0 h–0Si; (**b**) 72 h–0Si; (**c**) 144 h–0Si; (**d**) 216 h–0Si; (**e**) 288 h–0Si; (**f**) 360 h–0Si; (**g**) 72 h–0Si; (**h**) 72 h–0.8Si; (**i**) 144 h–0.8Si; (**j**) 216 h–0.8Si; (**k**) 288 h–0.8Si; (**l**) 360 h–0.8Si.

Because the Si content added to the low-density steel is relatively low, electron probe microanalysis (EPMA) was employed to accurately determine the distribution of Si within the rust layer by analyzing the elemental distribution across the rust-layer cross section. Based on the corrosion rate curves and macroscopic observations of the rust layers, relatively uniform corrosion products were formed after 360 h of periodic immersion, indicating that the corrosion process had entered a later stage; therefore, specimens subjected to 360 h of periodic immersion were selected for EPMA analysis. EPMA can qualitatively characterize the occurrence state of elements in the rust layer and the substrate. The color scale from blue to red represents the relative content of elements from low to high. By comparing the contrast of various colors in the figure can roughly determine where the elements are enriched. Figure 9 shows the elemental distributions in the rust layers of the 0Si steel after 360 h of periodic immersion, including both flat regions and pitting areas. It can be observed that, in the flat regions, the rust layer is mainly composed of Mn corrosion products, and an Fe-enriched zone is present at the interface between the rust layer and the substrate. The Al content in the rust layer is very low, with almost no Al-enriched regions, while the corrosive element Cl is relatively uniformly distributed throughout the entire rust layer. In contrast, in the pitting regions, there is almost no Mn enrichment inside the pits; therefore, the rust layer in these areas is considered to be mainly composed of corrosion products of Fe and Al. Moreover, a pronounced Al-enriched zone is observed near the substrate within the pits, and a large amount of Cl enrichment is also detected inside the pits.

https://doi.org/10.3390/ma19061182

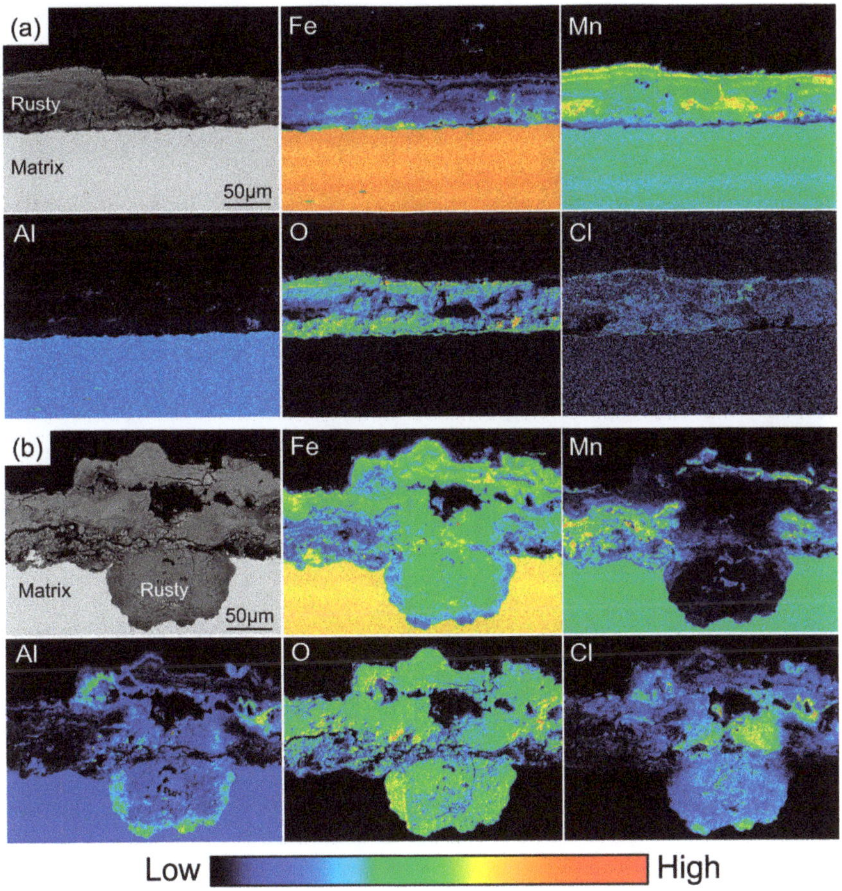


**Figure 9.** Elemental distribution of corrosion product cross-section in 0Si steel. (**a**) Flat region; (**b**) pitting region.

Figure 10 shows the elemental distributions in the rust layers of the 0.8Si steel after 360 h of periodic immersion, including both flat regions and pitting regions. In the flat regions, the rust layer is mainly composed of corrosion products of Mn and Fe; Fe is distributed within the interior of the rust layer, while Mn is heavily enriched in the outermost layer. Al-enriched zones are observed in both the outermost and innermost layers of the rust layer. O and Cl remain relatively uniformly distributed throughout the rust layer; however, a small amount of Cl enrichment is also detected at the Al-enriched positions in the outermost layer, indicating that Al plays a more pronounced role in hindering the inward diffusion of $Cl^-$ within the rust layer. A significant enrichment of Si is observed at the interface between the rust layer and the substrate, and this enrichment largely coincides with the Al-enriched regions within the rust layer. Combined with the XPS results, these Al- and Si-enriched regions are inferred to correspond to the surface oxide film that has just been dissolved. In the pitting regions, the rust layer inside the pits is still composed of corrosion products of Fe and Al, while Mn is distributed in the outermost layer of the entire rust layer. Si is also detected inside the pits, with its enrichment becoming more pronounced closer to the substrate, and Al remains enriched near the substrate.

https://doi.org/10.3390/ma19061182

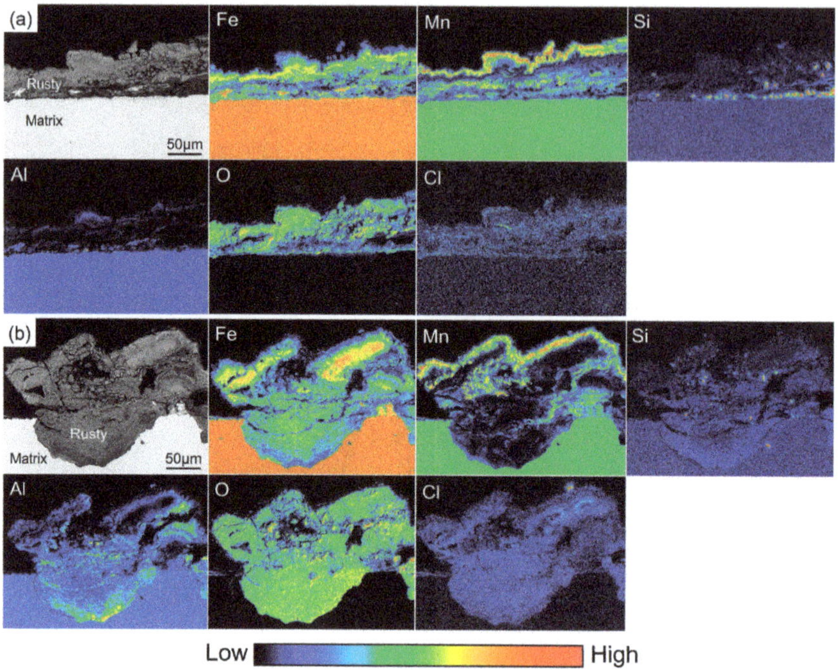


**Figure 10.** Elemental distribution of corrosion product cross-section in 0.8Si steel. (**a**) Flat region; (**b**) pitting region.

### 3.4.3. Phase Analysis of Corrosion Products

The differences in rust-layer morphology are essentially a macroscopic manifestation of variations in the phase composition of the corrosion products. To further determine the effect of Si on the corrosion products formed on the experimental steels after the periodic immersion, rust powders were collected from the steel surfaces and characterized by X-ray diffraction (XRD) to identify the rust-layer constituents and to perform semi-quantitative analysis of their contents. Figure 11 compares the main constituents of the rust layers formed on the experimental steels after 360 h of periodic immersion. Analysis of the XRD peaks indicates that the primary components of the rust layers are $Mn_3O_4$, α-FeOOH, γ-FeOOH, $Fe_2O_3$, $Fe_3O_4$, and $Al_2O_3$. After the addition of Si, the intensity and width of the diffraction peaks of some corrosion products change, indicating that Si influences the formation of certain oxides within the rust layer. No diffraction peaks associated with Si oxides are detected in the spectrum of the 0.8Si steel, which is likely due to their low content being below the detection limit of XRD [35].

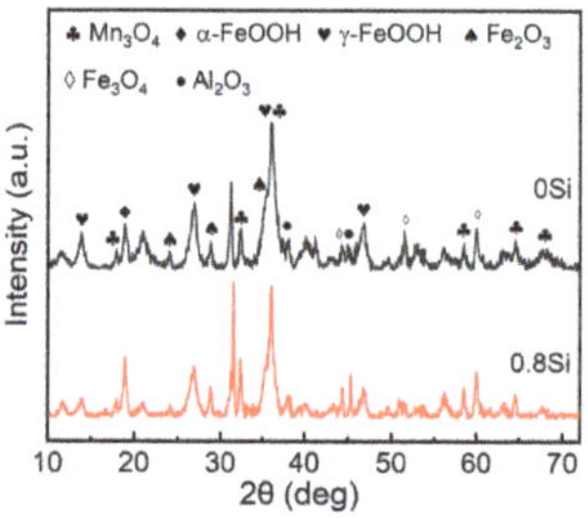


**Figure 11.** X-ray diffraction analysis of two steels after periodic immersion in 360 h.

https://doi.org/10.3390/ma19061182

Among the corrosion products of low-density steel, various iron oxides and oxyhydroxides remain the dominant constituents. Among them, γ-FeOOH is a metastable phase that can subsequently transform into the structurally stable α-FeOOH. The presence of $Fe_3O_4$ leads to an inhomogeneous distribution of corrosion products within the rust layer, and its porous nature prevents effective blockage of corrosive media ingress, thereby resulting in an increased corrosion rate at later stages. Therefore, to elucidate the effect of Si on the compactness of the rust layer in low-density steel and to provide an intuitive explanation of the influence of Si on the amount of corrosion products formed, a semi-quantitative calculation of the $\alpha/\gamma^*$ ratio was employed, where α represents α-FeOOH and $\gamma^*$ denotes the sum of γ-FeOOH and $Fe_3O_4$; the results are shown in Figure 12. After 360 h of exposure in the saline environment, the mass fraction of $Fe_3O_4$ in the rust layers of both experimental steels is identical, at 23%, while the mass fractions of γ-FeOOH are 66% and 58%, and those of α-FeOOH are 11% and 19%, respectively. After 360 h of periodic immersion, the rust layer of the 0.8Si steel exhibits a lower mass fraction of γ-FeOOH and a higher mass fraction of α-FeOOH, and the calculated $\alpha/\gamma^*$ value is higher than that of the 0Si steel. These results indicate that the addition of Si promotes the structural transformation of γ-FeOOH to α-FeOOH, increases the content of corrosion-resistant phases in the rust layer, and thereby enhances the compactness and stability of the rust layer, providing more effective protection for the substrate.

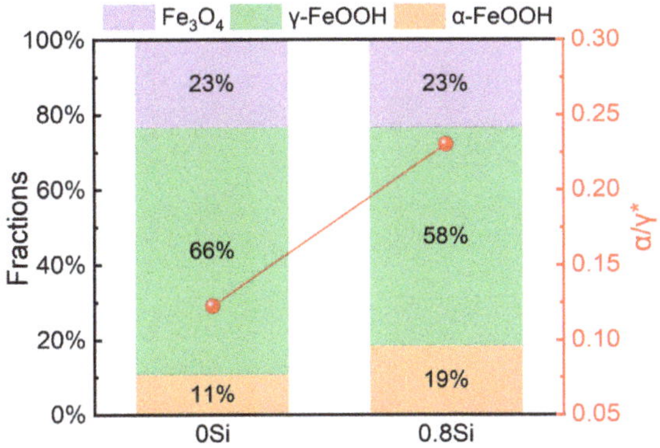


**Figure 12.** Proportion of each phase in rust layer of two steels after periodic immersion in 360 h. The red dots represent the $\alpha^*/\gamma$ ratio in samples with different Si contents.

### 3.4.4. Surface Morphology of Experimental Steels After Rust Removal

Due to the high Al content in the matrix of low-density steel, a dense surface oxide film forms rapidly during corrosion in a marine environment, resulting in pronounced passivation behavior. As corrosion progresses, defect sites in the oxide film are broken down by $Cl^-$, leading to the formation of pitting pits. These pits propagate into the depth of the material and gradually become crack initiation sites; the cracks continue to grow, ultimately resulting in material failure. Figure 13 shows the surface morphologies of the derusted specimens of the two experimental steels after periodic immersion, as captured by a three-dimensional laser confocal microscope. Overall, regardless of whether the corrosion duration is short (72 h) or long (360 h), the pits formed on the 0.8Si steel are smaller and shallower than those on the 0Si steel, which is consistent with the superior compactness of the oxide film on the 0.8Si steel. The maximum pit depths on the surfaces of the two experimental steels are shown in Figure 13e,f. After 72 h of periodic immersion, the pit depths of the two steels are similar, at 128 μm and 129 μm, respectively; however, the number and size of pits on the 0Si steel are significantly greater than those on the 0.8Si steel.

https://doi.org/10.3390/ma19061182

This indicates that the oxide films initially formed on both steels can be broken down by $Cl^-$, but the 0Si steel possesses more defect sites and a less compact oxide film, resulting in a greater number of pit initiation sites. After 360 h of periodic immersion, the maximum pit depth on the surface of the 0.8Si steel is 296 μm, which is markedly lower than that of the 0Si steel (349 μm), indicating that the growth rate of pit depth in the 0Si steel is significantly faster than that in the 0.8Si steel. This finding demonstrates that the addition of Si not only enhances the compactness of the oxide film formed at the initial stage of corrosion in low-density steel, but also facilitates its timely reconstruction after breakdown, thereby hindering further deepening of pitting pits. The EPMA results reveal pronounced enrichment of Al and Si at the bottoms of the pits, which is consistent with this conclusion.

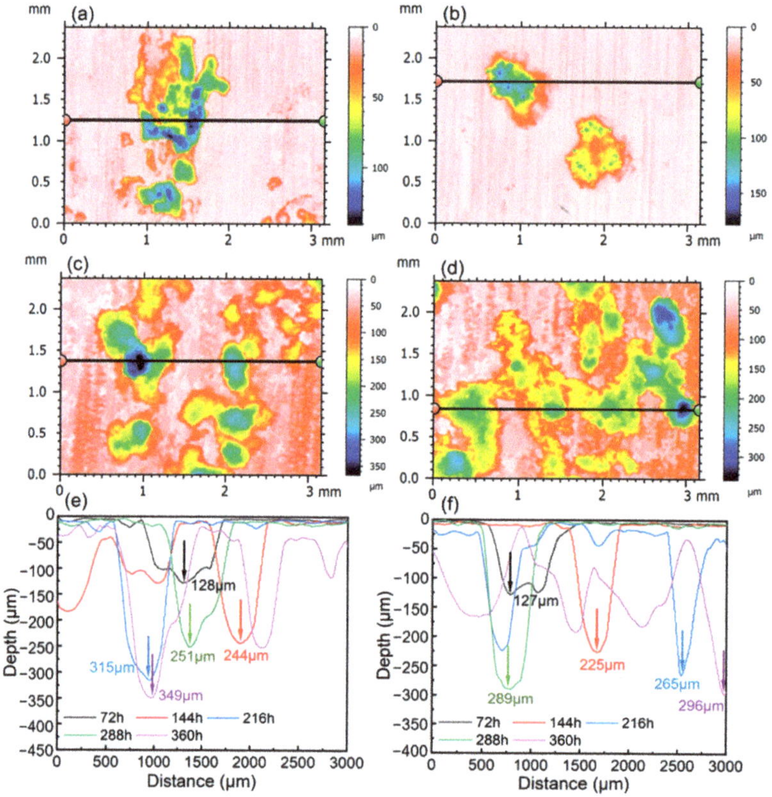


**Figure 13.** Surface morphology and maximum pit depth of two experimental steels. (**a**) Surface morphology of 0Si steel after 72 h; (**b**) Surface morphology of 0.8Si steel after 72 h; (**c**) Surface morphology of 0Si steel after 360 h; (**d**) Surface morphology of 0.8Si steel after 360 h; The maximum pit depth was measured along the horizontal line passing through the deepest pit point in (**a**–**d**). (**e**) Maximum pit depth of 0Si steel; (**f**) maximum pit depth of 0.8Si steel.

## 4. Discussion

### *4.1. Elemental Distribution in Oxide Films*

The relative contents of the elements in the oxide films of the two experimental steels were calculated using Equation (2) [36], and the results are presented in Figure 14.

$$C_X = \frac{I_X/S_X}{\sum I_i/S_i} \quad (2)$$

https://doi.org/10.3390/ma19061182

where $I_X$ is the peak area of the corresponding element and $S_X$ is the XPS sensitivity factor. Because low-density steel contains a relatively high Al content, Al is rapidly oxidized at the early stage of corrosion to form a dense alumina film; therefore, the dominant metallic element in the oxide film is expected to be Al. However, Fe is identified as the major metallic element in the oxide film of the 0Si steel, indicating that additional corrosion products have already formed outside the oxide film, which indirectly demonstrates the poor compactness of the oxide film. In contrast, Al is the dominant metallic element in the oxide film of the 0.8Si steel, confirming that only a dense oxide film is formed on the sample surface at this stage. In the 0.8Si steel, the relative contents of Si and Fe are approximately comparable; however, since the Si content in the steel matrix is much lower than that of Fe, it can be inferred that, during the initial stage of corrosion in a marine environment, Si reacts with oxygen more readily than Fe to form oxides. Both experimental steels exhibit the lowest Mn content in the oxide films, with the Mn content being even lower in the 0.8Si steel. During the initial corrosion process of austenitic low-density steel, Mn tends to form oxides with poor compactness, which deteriorate the compactness of the oxide film; therefore, this finding demonstrates that the addition of Si not only participates in oxidation to form $SiO_2$ but also significantly suppresses the oxidation of Mn and promotes the oxidation of Al, thereby enhancing the compactness of the oxide film. Comparable Al- and Si-enriched surface films in chloride-containing solutions have been reported for Fe-Mn-Al-Si steels, and these studies consistently show that the corrosion performance is highly sensitive to the balance between Al/Si-derived oxides and Mn-derived oxides/hydroxides in the surface film [37]. In particular, Mn enrichment in the outer oxide layer (e.g., $MnO/MnO_2$) has been associated with a less protective film and increased corrosion susceptibility, whereas $Al_2O_3$ and Si-containing ox-ides/hydroxides contribute to a denser and more stable surface layer [20].

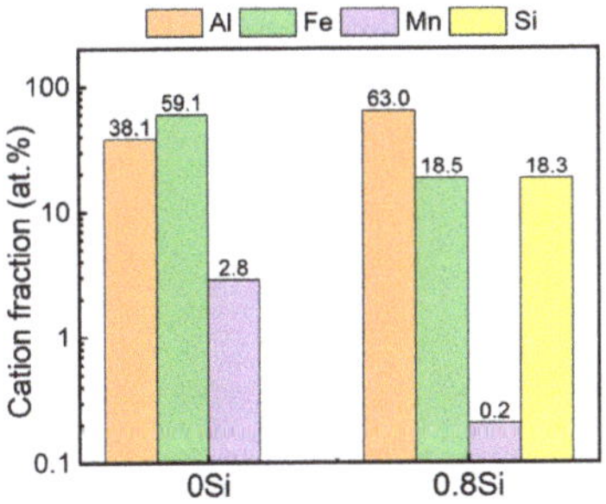


**Figure 14.** Variation in each element in the oxide film of the experimental steel.

### *4.2. Thermodynamic Analysis of Surface Oxide Film Formation*

The oxide film formed at the initial stage of corrosion plays a crucial role in determining the corrosion resistance of austenitic low-density steel. In the oxide film, Al is the dominant metallic element, while Si also undergoes rapid oxidation and forms $SiO_2$ within the oxide film. The standard Gibbs free energy ($\Delta G^\circ$) values of the oxidation reactions of Al, Si, Fe, and Mn at room temperature are listed in Table 4. According to the Ellingham diagram, a lower standard Gibbs free energy corresponds to a stronger tendency for oxidation. The oxidation curve of Al consistently appears at the bottom of the Ellingham diagram, indicating that Al preferentially oxidizes to form thermodynamically stable $Al_2O_3$ even under oxygen-deficient conditions. The oxidation curve of Si lies only slightly above that of Al in the Ellingham diagram and thus Si is also readily oxidized [38]; however, due to its very low content, the XPS results of the 0.8Si steel shown in Figure 14 indicate that, at the early stage of corrosion, the content of $SiO_2$ in the oxide film is lower than that of

https://doi.org/10.3390/ma19061182

$Al_2O_3$. In contrast, the oxidation reactions of Fe and Mn exhibit higher $\Delta G^\circ$ values, making them less prone to oxidation under oxygen-deficient conditions [39]; therefore, relatively lower fractions of Fe and Mn are observed in the phase analysis of the oxide film. From an electrochemical perspective, the standard electrode potentials ($E^\circ$) of these four elements are listed in Table 5. The more negative the standard electrode potential, the stronger the tendency of a metal to undergo oxidation. Al exhibits the most negative potential, demonstrating its strongest tendency for electron loss and anodic dissolution, which provides the electrochemical driving force for its oxidation process. Therefore, the present thermodynamic analysis should be interpreted together with kinetic and transport effects: in NaCl solutions, selective oxidation, competitive dissolution, and element redistribution can lead to an outer Mn-rich, defective layer and an inner Al-rich layer, as reported by XPS and EIS studies on high-Mn austenitic Fe-Mn-Al(-Si) steels. This framework is consistent with Figure 14, where Si addition reduces Mn incorporation into the oxide film while maintaining an Al-dominant film chemistry, which is expected to increase film resistance and retard chloride-assisted breakdown [37].

**Table 4.** Standard Gibbs free energy of major alloying elements at room temperature.

| | $Al \rightarrow Al_2O_3$ | $Fe \rightarrow FeO$ | $Mn \rightarrow MnO$ | $Si \rightarrow SiO_2$ |
|---|---|---|---|---|
| $\Delta G^\circ$ (kJ/mol) | −756.3 | −116.7 | −173.4 | −856.6 |

**Table 5.** Standard electrode potentials of major alloying elements at room temperature.

| | $Al^{3+}/Al$ | $Mn^{2+}/Mn$ | $Fe^{2+}/Fe$ | $SiO_2/Si$ |
|---|---|---|---|---|
| $E^\circ$ ($V_{vs.SCE}$) | −1.66 | −1.18 | −0.44 | −0.86 |

### *4.3. Transformation Mechanism of Corrosion Products*

During the corrosion process of low-density steel, an oxide film preferentially forms on the substrate surface. As the exposure time increases, the oxide film gradually dissolves, substrate elements progressively diffuse outward, and a brown–black rust layer, as shown in Figure 8, is ultimately formed. Because Al has the most negative standard Gibbs free energy for oxide formation, it exhibits the fastest outward diffusion rate and therefore appears in the outermost layer of the rust layer, followed by Mn, while Fe is mainly distributed in the innermost layer. However, no Al is observed in the rust layer of the 0Si steel shown in Figure 9, whereas partial Al enrichment is detected in the outermost layer of the rust layer of the 0.8Si steel shown in Figure 10. This indicates that Al is indeed distributed in the outermost rust layer but tends to combine with $Cl^-$ and subsequently spall off, which also explains why relatively low Cl enrichment is observed in the rust layers at the flat regions in Figures 9 and 10. In the innermost region of the flat rust layer of the 0.8Si steel, Al- and Si-enriched zones are detected, which are considered to correspond to the oxide film that has just been dissolved. This observation demonstrates that the addition of Si enhances the stability of $Al_2O_3$, resulting in a more compact oxide film that is less prone to dissolution compared with that formed on the 0Si steel. In addition to dissolution, the oxide film can also be broken down by $Cl^-$ ions, leading to the formation of pitting pits; the enrichment of Al and Si at the bottoms of the pits indicates that the oxide film undergoes reconstruction after breakdown, thereby hindering further deepening of the pits. Consequently, as shown in Figure 13, the 0.8Si steel exhibits a lower pitting depth. Similar synergistic effects between Al- and Si-containing oxides/hydroxides have been reported for Fe-Mn-Al-Si-C alloys in NaCl media, where hydrated alumina and amor-phous silica

https://doi.org/10.3390/ma19061182

species form a relatively protective scale, while Mn-rich compounds are generally less compact [40].

Studies have shown that the chemical composition and compactness of the rust layer determine its ability to inhibit corrosion of the substrate, and that the relative proportions of oxidation products such as α-FeOOH, β-FeOOH, γ-FeOOH, $Fe_3O_4$, and $Fe_2O_3$ can differ significantly under identical service conditions [41]. Among the iron oxyhydroxides present in rust layers, the corrosion resistance increases in the order β-FeOOH < γ-FeOOH < α-FeOOH. In marine environments containing high concentrations of $Cl^-$, large amounts of β-FeOOH are generally formed [42]. β-FeOOH possesses a relatively loose structure, is prone to moisture absorption, and exhibits strong chloride-ion adsorption capability and oxidizing activity. However, β-FeOOH is not detected in the XRD patterns shown in Figure 11. This is because β-FeOOH, as a highly reactive and readily reducible intermediate product, although it forms in large quantities within the rust layer, has largely transformed into more stable oxides such as $Fe_3O_4$ and α-FeOOH after 360 h of periodic immersion, and therefore its diffraction peaks are not observed in XRD measurements [43]. The reduction priority of γ-FeOOH is lower than that of β-FeOOH; nevertheless, this metastable phase ultimately transforms into the structurally stable α-FeOOH. α-FeOOH has an orthorhombic crystal structure, and owing to its high chemical stability, it is generally regarded as a protective corrosion product in rust layers. Under specific environmental conditions, α-FeOOH can effectively inhibit further corrosion of the substrate [44], thereby slowing the corrosion rate of low-density steel and providing protection to the substrate. In addition, immersion and polarization studies on Fe-Mn-Al-Si steels in 3.5% NaCl indicate that corrosion tends to be localized and strongly governed by passive-film compactness and its resistance to chloride-induced breakdown, supporting the emphasis on film integrity and reconstruction in the present work.

## 5. Conclusions

Electrochemical experiments conducted in a 3.5 wt.% NaCl solution yielded the potentiodynamic polarization curves and electrochemical impedance spectra of the two experimental steels. Compared with the 0Si steel ($E_{corr}$ = −0.93 ± 0.024 $V_{vs.SCE}$, $I_{corr}$ = 4.28 ± 1.233 μA/cm$^2$, $E_p$ = −0.48 ± 0.018 $V_{vs.SCE}$, $I_p$ = 27.79 ± 2.563 μA/cm$^2$), the 0.8Si steel shows a nobler $E_{corr}$ (−0.90 ± 0.020 $V_{vs.SCE}$) and higher $E_p$ (−0.38 ± 0.016 $V_{vs.SCE}$) together with reduced $I_{corr}$ (3.84 ± 0.985 μA/cm$^2$) and $I_p$ (20.94 ± 3.070 μA/cm$^2$). Mean-while, the 0.8Si steel exhibits a larger impedance arc radius with three time-constant features, indicating higher charge-transfer/film resistance and thus superior corrosion resistance.

Potentiostatic polarization tests performed on both experimental steels resulted in the formation of dense oxide films on the surfaces. Within the first ~200 s of polarization, the current density rapidly decreased to ~3 μA/cm$^2$, and the 0.8Si steel reached a lower steady-state current density after 7200 s at −0.5 $V_{SCE}$. Cross-sectional TEM reveals that the oxide film on the 0Si steel is highly nonuniform (≈75–127 nm), whereas the 0.8Si steel forms a thicker and much more uniform film (≈138 nm). Consistently, XPS shows FeOOH is detected in the 0Si film but is absent in the 0.8Si film, while Si participates in film formation as $SiO_2$ (Si 2p at 103.4 eV) together with $Al_2O_3$ (Al 2p at 74.3 eV), demonstrating that Si addition improves oxide-film compactness/uniformity, enhances substrate protection, and reduces pitting initiation sites.

Periodic immersion accelerated corrosion tests on the two low-density steels reveal that the corrosion rate is significantly reduced after the addition of Si throughout 72–360 h exposure. After 360 h, XRD semi-quantification indicates identical $Fe_3O_4$ mass fractions (23%) for both steels, but γ-FeOOH decreases from 66% (0Si) to 58% (0.8Si) while α-FeOOH

https://doi.org/10.3390/ma19061182

increases from 11% to 19%, resulting in a higher $\alpha/\gamma^*$ ratio and a more stable/compact rust layer. EPMA analysis shows pronounced Al and Si enrichment in the inner rust layer and at pitting cavities, and 3D confocal measurements further con-firm suppressed pit growth: the maximum pit depth is reduced from 349 μm (0Si) to 296 μm (0.8Si) after 360 h (72 h: 128 μm vs. 129 μm). Together, these results demonstrate that Si effectively inhibits the initiation and propagation of pitting corrosion by promoting protective film/rust-layer evolution.

**Author Contributions:** Conceptualization, S.W., Y.H. and J.L.; methodology, J.L., Y.H. and J.Q.; validation, J.Q., S.W., H.W., Y.H. and J.L.; formal analysis, J.Q.; investigation, J.Q. and J.L.; resources, J.Q., H.W. and S.W.; data curation, J.Q. and Y.H.; writing—original draft preparation, Y.H., J.Q., H.W. and J.L.; writing—review and editing, S.W. and Y.H.; visualization, J.Q.; supervision, J.L.; project administration, S.W.; funding acquisition, H.W. All authors have read and agreed to the published version of the manuscript.

**Funding:** This research was funded by Advanced Materials-National Science and Technology Major Project (2024ZD0601600).

**Data Availability Statement:** The original contributions presented in this study are included in the article. Further inquiries can be directed to the corresponding author.

**Conflicts of Interest:** The authors declare no conflicts of interest.

## References

1. Shojai, S.; Schaumann, P.; Braun, M.; Ehlers, S. Influence of pitting corrosion on the fatigue strength of offshore steel structures based on 3D surface scans. *Int. J. Fatigue* **2022**, *164*, 107128. [CrossRef]
2. Zanuttigh, B.; Dallavalle, E.; Zagonari, F. A novel framework for sustainable decision-making on reusing Oil & Gas offshore platforms with application to the Adriatic Sea. *Renew. Sustain. Energy Rev.* **2025**, *211*, 115252.
3. Karimi, H.R.; Shabakhty, N. A comparative assessment of the impact of tilting on structural safety and behavior of fixed offshore platforms. *Appl. Ocean Res.* **2023**, *136*, 103589. [CrossRef]
4. Edwards, E.C.; Holcombe, A.; Brown, S.; Ransley, E.; Hann, M.; Greaves, D. Trends in floating offshore wind platforms: A review of early-stage devices. *Renew. Sustain. Energy Rev.* **2024**, *193*, 114271. [CrossRef]
5. Ben, N.; Vytyaz, O.Y.; Hrabovskyy, R.S. Mechanical Properties of Steel for Floating Offshore Platforms Under Static and Cyclic Loading. *Mater. Sci.* **2023**, *59*, 152–157. [CrossRef]
6. Mimica, D.; Skejic, D.; Catipovic, I.; Parunov, J. Framework for the Lifetime Extension of Fixed Offshore Platforms in the Northern Adriatic Sea. In Proceedings of the 26th Symposium on the Theory and Practice of Shipbuilding-SORTA, Zadar, Croatia, 2–5 October 2024; pp. 151–167.
7. Shojai, S.; Schaumann, P.; Bromer, T. Probabilistic modelling of pitting corrosion and its impact on stress concentrations in steel structures in the offshore wind energy. *Mar. Struct.* **2022**, *84*, 103232. [CrossRef]
8. Liu, C.; Xu, Q.; Liu, X.; Fan, X.; Li, Y. Comparative study of marine steel and welding joint in artificial seawater based on stress corrosion cracking and crack growth. *J. Mater. Res. Technol.* **2024**, *33*, 2683–2691. [CrossRef]
9. Amin, Z.; Hamed, K.; Afshin, R. Fault-tolerant control for nonlinear offshore steel jacket platforms based on reinforcement learning. *Ocean Eng.* **2022**, *246*, 110247. [CrossRef]
10. Ji, H.; Wang, H.; Chen, Q.; Ma, X.; Cai, Y. Corrosion behavior prediction for hull steels under dynamic marine environments by jointly utilizing LSTM network and PSO-RF model. *Ocean Eng.* **2024**, *300*, 117371. [CrossRef]
11. Lu, Y.; Liu, Z.X.; Wang, P.P.; Wang, H.Y. Corrosion failure mechanism and electrolytic accelerated fatigue performance of butt-welded joints. *J. Constr. Steel Res.* **2025**, *227*, 109354. [CrossRef]
12. Liu, H.X.; Gao, H.M.; Chen, J.M.; Liu, R.L.; Zhang, Y.; Yin, Y.S.; Liu, H.F.; Fan, S.J.; Liu, H.W. Study of corrosion and failure mechanism of the galvanized steel pipes of meteorological tower serviced in the South China Sea. *Eng. Fail. Anal.* **2025**, *167*, 108933. [CrossRef]
13. Du, J.; Chen, P.; Zhang, F.; Jia, Z.; Shi, F.; Li, X. Controllable κ-carbide precipitation enables strength-ductility co-enhancement in Fe-Mn-Al-C low-density austenitic steel via grain boundary engineering. *J. Mater. Sci. Technol.* **2025**, *227*, 26–31. [CrossRef]
14. Wang, S.T.; Sun, Z.D.; Li, D.J.; Yu, Q.; Wang, Q.F. Effect of Aluminum Content on the Corrosion Behavior of Fe-Mn-Al-C Structural Steels in Marine Environments. *Metals* **2025**, *15*, 1249. [CrossRef]

https://doi.org/10.3390/ma19061182

15. Gutierrez-Urrutia, I. Low Density Fe-Mn-Al-C Steels: Phase Structures, Mechanisms and Properties. *ISIJ Int.* **2021**, *61*, 16–25. [CrossRef]
16. Zhao, L.; van Dijk, N.H.; Brück, E.; Sietsma, J.; van der Zwaag, S. Magnetic and X-ray diffraction measurements for the determination of retained austenite in TRIP steels. *Mater. Sci. Eng. A-Struct. Mater. Prop. Microstruct. Process.* **2001**, *313*, 145–152. [CrossRef]
17. Bhattacharya, B.; Sharma, A.S.; Hazra, S.S.; Ray, R.K. A Study of Microstructures and Tensile Properties of Two Fe-Mn-Al-Si-C Alloys. *Metall. Mater. Trans. A-Phys. Metall. Mater. Sci.* **2009**, *40A*, 1190–1202. [CrossRef]
18. Jia, Q.J.; Jiang, X.Q.; Wu, C.J.; Chen, J.X.; Zhu, X.Y.; Liu, Y.; Su, X.P. Effect of Si on Mechanical Properties and Oxide Film Formation of AFA Alloy at Low Oxygen Pressure. *Coatings* **2025**, *15*, 602. [CrossRef]
19. Heon-Young, H.; Kyeong-Won, K.; Seong-Jun, P.; Tae-Ho, L.; Hyungkwon, P.; Joonoh, M.; Hyun-Uk, H.; Chang-Hoon, L. Effects of Cr on pitting corrosion resistance and passive film properties of austenitic Fe–19Mn–12Al–1.5 C lightweight steel. *Corros. Sci.* **2022**, *206*, 110529.
20. Bosch, J.; Martin, U.; Aperador, W.; Bastidas, J.M.; Ress, J.; Bastidas, D.M. Corrosion Behavior of High-Mn Austenitic Fe–Mn–Al–Cr–C Steels in NaCl and NaOH Solutions. *Materials* **2021**, *14*, 425. [CrossRef]
21. Jinbin, Z.; Wei, W.; Jiaxing, C.; Xuequn, C. Evaluating the effect of aluminum on the corrosion resistance of the structural steels used for marine engineering. *J. Mater. Res. Technol.* **2022**, *18*, 4181–4193. [CrossRef]
22. Haixia, L.; Feng, H.; Wei, Y.; Qian, H.; Jing, L.; Frank, C.Y. Essential role of element Si in corrosion resistance of a bridge steel in chloride atmosphere. *Corros. Sci.* **2020**, *173*, 108758. [CrossRef]
23. Liu, Y.; Pan, H.; Zhao, Y.; Zhou, L.; Feng, J.; Jiang, Y. New insights for composition design: A novel synergistic corrosion-resistant effect of Fe–Mn–Al–Cr–Si–Mo–C lightweight steel in 3.5 wt% NaCl solution. *J. Mater. Res. Technol.* **2024**, *32*, 1602–1619. [CrossRef]
24. *GB/T 228.1-2021*; Metallic Materials—Tensile Testing—Part 1: Method of Test at Room Temperature. Standards Press of China: Beijing, China, 2021.
25. *GB/T 229-2020*; Metallic Materials—Charpy Pendulum Impact Test Method. Standards Press of China: Beijing, China, 2020.
26. Kozeschnik, E.; Bhadeshia, H.K.D.H. Influence of silicon on cementite precipitation in steels. *Mater. Sci. Technol.* **2008**, *24*, 343–347. [CrossRef]
27. Mingjie, S.; Baojian, H.; Tao, Y. Micromechanical simulations and experimental characteristics of randomly distributed carbon nanotubes reinforced Mg matrix composites. *J. Alloys Compd.* **2022**, *924*, 166589.
28. Sarac, B.; Sharifikolouei, E.; Zheng, Y.H.; Yüce, E.; Asci, A.; Keckes, J.; Sarac, A.S.; Eckert, J. Electrochemical impedance behavior and corrosion resistance of amorphous 316-type stainless steel microfibers in saline environment. *Mater. Today Commun.* **2025**, *44*, 112178. [CrossRef]
29. Zhao, Q.C.; Luo, H.; Li, C.T.; Yan, B.B.; Liang, G.X. Examining the passive properties of laser powder bed fused austenitic, stainless steel in hydrochloric acid: Effect of building directions. *Constr. Build. Mater.* **2026**, *506*, 144943. [CrossRef]
30. Wang, D.S.; Sun, X.H.; Liu, W.; Wang, B.; Sun, S.B.; Jiang, Y.C.; Chang, X.T. Effect of ammonium ions on the corrosion behaviors of high-manganese austenitic steel. *Int. J. Electrochem. Sci.* **2025**, *20*, 101170. [CrossRef]
31. Zhou, Y.L.; Chen, J.; Liu, Z.Y. Corrosion Behavior of Rusted 550 MPa Grade Offshore Platform Steel. *J. Iron Steel Res. Int.* **2013**, *20*, 66–73. [CrossRef]
32. Refait, P.; Jeannin, M.; Urios, T.; Fagot, A.; Sabot, R. Corrosion of low alloy steel in stagnant artificial or stirred natural seawater: The role of Al and Cr. *Mater. Corros.* **2019**, *70*, 985–995. [CrossRef]
33. Zhang, D.; Zhang, X.; Wang, S.; Zhang, F.; Ren, H.; Zhang, J.; Yan, Q. A comparative study on the corrosion resistance and microstructure stability of Fe-Ni-Cr-Al-Si based austenitic steel in lead-bismuth eutectic with $10^{-6}$ wt.% oxygen at 600 and 650 °C for up to 6000 h. *J. Nucl. Mater.* **2026**, *619*, 156295. [CrossRef]
34. Dai, C.; Luo, H.; Li, J.; Du, C.W.; Liu, Z.Y.; Yao, J.Z. X-ray photoelectron spectroscopy and electrochemical investigation of the passive behavior of high-entropy FeCoCrNiMox alloys in sulfuric acid. *Appl. Surf. Sci.* **2020**, *499*, 143903. [CrossRef]
35. Kimura, M.; Kihira, H.; Ohta, N.; Hashimoto, M.; Senuma, T. Control of $Fe(O,OH)_6$ nano-network structures of rust for high atmospheric-corrosion resistance. *Corros. Sci.* **2005**, *47*, 2499–2509. [CrossRef]
36. Yang, G.; Wang, L.; Fu, X.Q.; Ji, Y.C.; Cao, W.Q.; Wang, C.Y.; Liang, J.X.; Dong, C.F. Enhancing the corrosion resistance of low-density steels via Al-rich oxide film formation by grain size optimization. *Corros. Sci.* **2025**, *257*, 113331. [CrossRef]
37. Fajardo, S.; Llorente, I.; Jiménez, J.A.; Bastidas, J.M.; Bastidas, D.M. Effect of Mn additions on the corrosion behaviour of TWIP Fe-Mn-Al-Si austenitic steel in chloride solution. *Corros. Sci.* **2019**, *154*, 246–253. [CrossRef]
38. Zhang, L.B.; Ning, F.Q.; Yan, H.; Liu, J.; Wang, H. The effect of Si on the microstructure and corrosion resistance of XSi9Cr ferritic/martensitic steel in oxygen-saturated lead-bismuth eutectic. *Corros. Sci.* **2025**, *255*, 113079. [CrossRef]
39. Eijk, C.v.d.; Dalaker, H.; Safarian, J. Possibilities and Limitations of the Use of Hydrogen in Different Metallurgical Sectors. *Mater. Proc.* **2023**, *15*, 63.
40. Lins, V.F.C.; Freitas, M.A.; Silva, E.M.P.E. Corrosion resistance study of Fe–Mn–Al–C alloys using immersion and potentiostatic tests. *Appl. Surf. Sci.* **2004**, *250*, 124–134. [CrossRef]


https://doi.org/10.3390/ma19061182

41. Monnier, J.; Burger, E.; Berger, P.; Neff, D.; Guillot, I.; Dillmann, P. Localisation of oxygen reduction sites in the case of iron long term atmospheric corrosion. *Corros. Sci.* **2011**, *53*, 2468–2473. [CrossRef]
42. Chinara, M.; Ghosh, R.; Mukherjee, S.; Mondal, K. Corrosion mechanism of line pipe steels (API X70 and X80 grades) under aggressive salt-spray exposure. *Mater. Chem. Phys.* **2025**, *339*, 130762. [CrossRef]
43. Li, Z.L.; Song, J.L.; Chen, J.H.; Yu, Q.; Xiao, K. Corrosion behavior of a high-strength steel E690 in aqueous electrolytes with different chloride concentrations. *J. Mater. Res. Technol.* **2023**, *22*, 596–604. [CrossRef]
44. Melchers, R.E. Effect of small compositional changes on marine immersion corrosion of low alloy steels. *Corros. Sci.* **2004**, *46*, 1669–1691. [CrossRef]



https://doi.org/10.3390/ma19061182

MDPI AG
Grosspeteranlage 5
4052 Basel
Switzerland
Tel.: +41 61 683 77 34

*Materials* Editorial Office
E-mail: materials@mdpi.com
www.mdpi.com/journal/materials

www.ingramcontent.com/pod-product-compliance
Lightning Source LLC
LaVergne TN
LVHW070449061126
851858LV00011B/193